TUTORIAL CHEMISTRY TEXTS

18

Maths for Chemists

Volume 1

Numbers, Functions and Calculus

MARTIN COCKETT & GRAHAM DOGGETT

University of York

RS•C

ROYAL SOCIETY OF CHEMISTRY

ISBN 0-85404-677-1

A catalogue record for this book is available from the British Library

Published by The Royal Society of Chemistry, Thomas Graham House, Science Park,
Milton Road, Cambridge CB4 0WF, UK
Registered Charity No. 207890
For further information see our web site at www.rsc.org

Typeset in Great Britain by Alden Bookset, Northampton
Printed and bound in Italy by Rotolito Lombarda

Preface

These two introductory texts provide a sound foundation in the key mathematical topics required for degree level chemistry courses. While they are primarily aimed at students with limited backgrounds in mathematics, the texts should prove accessible and useful to all chemistry undergraduates. We have chosen from the outset to place the mathematics in a chemical context – a challenging approach because the context can often make the problem appear more difficult than it actually is. However, it is equally important to convince students of the relevance of mathematics in all branches of chemistry. Our approach links the key mathematical principles with the chemical context by introducing the basic concepts first, and then demonstrates how they translate into a chemical setting.

Historically, physical chemistry has been the target for mathematical support; however, in all branches of chemistry – be they the more traditional areas of inorganic, organic and physical, or the newer areas of biochemistry, analytical and environmental chemistry – mathematical tools are required to build models of varying degrees of complexity, in order to develop a language for providing insight and understanding together with, ideally, some predictive capability.

Since the target student readership possesses a wide range of mathematical experience, we have created a course of study in which selected key topics are treated without going too far into the finer mathematical details. The first two chapters of Volume 1 focus on numbers, algebra and functions in some detail, as these topics form an important foundation for further mathematical developments in calculus, and for working with quantitative models in chemistry. There then follow chapters on limits, differential calculus, differentials and integral calculus. Volume 2 covers power series, complex numbers, and the properties and applications of determinants, matrices and vectors. We avoid discussing the statistical treatment of error analysis, in part because of the limitations imposed by the format of this series of tutorial texts, but also because the procedures used in the processing of experimental results are commonly provided by departments of chemistry as part of their programme of practical chemistry courses. However, the propagation of errors, resulting from the use of formulae, forms part of the chapter on differentials in Volume 1.

Martin Cockett
Graham Doggett
York

TUTORIAL CHEMISTRY TEXTS

EDITOR-IN-CHIEF

Professor E W Abel

EXECUTIVE EDITORS

Professor A G Davies
Professor D Phillips
Professor J D Woollins

EDUCATIONAL CONSULTANT

Mr M Berry

This series of books consists of short, single-topic or modular texts, concentrating on the fundamental areas of chemistry taught in undergraduate science courses. Each book provides a concise account of the basic principles underlying a given subject, embodying an independent-learning philosophy and including worked examples. The one topic, one book approach ensures that the series is adaptable to chemistry courses across a variety of institutions.

TITLES IN THE SERIES

Stereochemistry *D G Morris*
Reactions and Characterization of Solids
 S E Dann
Main Group Chemistry *W Henderson*
d- and f-Block Chemistry *C J Jones*
Structure and Bonding *J Barrett*
Functional Group Chemistry *J R Hanson*
Organotransition Metal Chemistry *A F Hill*
Heterocyclic Chemistry *M Sainsbury*
Atomic Structure and Periodicity *J Barrett*
Thermodynamics and Statistical Mechanics
 J M Seddon and J D Gale
Basic Atomic and Molecular Spectroscopy
 J M Hollas
Organic Synthetic Methods *J R Hanson*
Aromatic Chemistry *J D Hepworth,*
 D R Waring and M J Waring
Quantum Mechanics for Chemists
 D O Hayward
Peptides and Proteins *S Doonan*
Reaction Kinetics *M Robson Wright*
Natural Products: The Secondary
 Metabolites *J R Hanson*
Maths for Chemists, Volume 1, Numbers,
Functions and Calculus *M Cockett and*
G Doggett
Maths for Chemists, Volume 2, Power Series,
Complex Numbers and Linear Algebra
M Cockett and G Doggett

FORTHCOMING TITLES

Mechanisms in Organic Reactions
Molecular Interactions
Lanthanide and Actinide Elements
Bioinorganic Chemistry
Chemistry of Solid Surfaces
Biology for Chemists
Multi-element NMR
EPR Spectroscopy
Biophysical Chemistry

Further information about this series is available at www.rsc.org/tct

Order and enquiries should be sent to:
Sales and Customer Care, Royal Society of Chemistry, Thomas Graham House, Science Park, Milton Road, Cambridge CB4 0WF, UK

Tel: +44 1223 432360; Fax: +44 1223 426017; Email: sales@rsc.org

Contents

Symbols

>	greater than	∝	proportionality
≥	greater than or equal to	=	equality
≫	much greater than	∞	infinity
<	less than	Σ	summation sign
≤	less than or equal to		
≪	much less than	Π	product sign
/ or ÷	division	!	factorial
≠	not equal to	{}	braces
≅ or ≈	approximately equal to	[ ]	brackets
⇒	implies	( )	parentheses
∈	element of		

1

Numbers and Algebra

Numbers of one kind or another permeate all branches of chemistry (and science generally), simply because any measuring device we use to record a characteristic of a system can only yield a number as output. For example, we might measure or determine the:

- Weight of a sample.
- Intensity or frequency of light absorption of a solution.
- Vibration frequency for the HCl molecule.
- Relative molecular mass of a carbohydrate molecule.

Or we might:

- Confirm the identity of an organic species by measuring its boiling point.
- Measure, or deduce, the equilibrium constant of a reversible reaction.
- Wish to count the number of isomeric hydrocarbon species with formula C_4H_{10}.

In some of these examples, we also need to:

- Specify units.
- Estimate the error in the measured property.

Clearly then, the manner in which we interact with the world around us leads us quite naturally to use numbers to interpret our experiences.

In many situations, we routinely handle very large and very small numbers, so disparate in size that it is difficult to have an intuitive feel for order of magnitude. For example:

- The number of coulombs (the basic unit of electrical charge) associated with a single electron is approximately 0.000 000 000 000 000 000 160 2177.
- The equilibrium constant for the electrochemical process

$$Au^{3+}(aq) + Al(s) \rightleftharpoons Au(s) + Al^{3+}(aq)$$

is of the order of 1 followed by 4343 zeros.[1] In chemical terms, we have no problem with this answer, as it indicates that the equilibrium is totally

Decimal numbers are commonly written with a space between every group of three digits after the decimal point (sometimes omitted if there are only four such digits).

towards the right side (which means that the aluminium electrode will be completely consumed and the gold electrode untouched).

These two widely different examples, of a type commonly experienced in chemistry, illustrate why it is so important to feel at ease using numbers of all types and sizes. A familiarity and confidence with numbers is of such fundamental importance in solving quantitative chemical problems that we devote the first two chapters of this book to underpinning these foundations. Our main objective is to supply the necessary tools for constructing models to help in interpreting numerical data, as well as in achieving an understanding of the significance of such data.

Aims

In this introductory chapter, we provide the necessary tools for working with numbers and algebraic symbols, as a necessary prelude to understanding functions and their properties – a key topic of mathematics that impinges directly on all areas of chemistry. By the end of the chapter you should be able to:

- Understand the different types of numbers and the rules for their combination
- Work with the scientific notation for dealing with very large and very small numbers
- Work with numerical and algebraic expressions
- Simplify algebraic expressions by eliminating common factors
- Combine rational expressions by using a common denominator
- Treat units as algebraic entities

1.1   Real Numbers

1.1.1   Integers

Counting numbers have been in use for a very long time, but the recognition of zero as a numeral originated in India over two millennia ago, and only became widely accepted in the West with the advent of the printed book in the 13th century (for further details, see D. Wells, *The Penguin Dictionary of Curious and Interesting Numbers*, Penguin, London, 1987).

One of the earliest skills we learn from childhood is the concept of counting: at first we learn to deal with **natural numbers** (positive, whole numbers), including zero, but we tend to ignore the concept of negative numbers, because they are not generally used to count objects. However, we soon run into difficulties when we have to subtract two numbers, as this process sometimes yields a negative result. The concept of a negative counting number applied to an object can lead us into all sorts of trouble, although it does allow us to account for the notion of debt (you owe me 2 apples is the equivalent of saying "I own -2 apples"). We therefore extend natural numbers to a wider category of number called **integers**, which consist of all positive and negative whole numbers, as well as zero:

$$\ldots, -3, -2, -1, 0, 1, 2, 3, \ldots$$

We use integers in chemistry to specify:

- The atomic number, Z, defined as the number of protons in the nucleus; Z is a positive integer, less than or equal to 109.
- The number of atoms of a given type (positive) in the formula of a chemical species.
- The number of electrons (a positive integer) involved in a redox reaction occurring in an electrochemical cell.
- The quantum numbers required in the mathematical specification of individual atomic orbitals. These can take positive or negative integer values or zero, depending on the choice of orbital.

At the time of writing, the heaviest element to have been isolated is the highly radioactive element meitnerium ($Z = 109$), of which only a few atoms have ever been made. The heaviest naturally occurring element is uranium, $Z = 92$.

1.1.2 Rational Numbers

When we divide one integer by another, we sometimes obtain another integer: For example, $6/-3 = -2$; at other times, however, we obtain a fraction, or **rational number**, of the form $\frac{a}{b}$, where the integers a and b are known as the **numerator** and **denominator**, respectively, for example, $\frac{2}{3}$. The denominator, b, cannot take the value zero because $\frac{a}{0}$ is of indeterminate value.

Rational numbers occur in chemistry:

- In defining the spin quantum number of an electron ($s = 1/2$), and the nuclear spin quantum number, I, of an atomic nucleus; for example, ^{45}Sc has $I = \frac{7}{2}$.
- In specifying the coordinates $(0,0,0)$ and $(\frac{a}{2}, \frac{a}{2}, \frac{a}{2})$, which define the locations of two of the nuclei that generate a body-centred unit cell of side a.

1.1.3 Irrational Numbers

Rational numbers can always be expressed as ratios of integers, but sometimes we encounter numbers which cannot be written in this form. These numbers are known as **irrational numbers** and include:

- **Surds**, of the form $\sqrt{2}, \sqrt[3]{2}$, which are obtained from the solution of a quadratic or higher order equation.
- **Transcendental numbers**, which, in contrast to surds, do not derive from the solution to algebraic equations. Examples include π, which

$\sqrt{2}$ is obtained as a solution of the equation $x^2 - 2 = 0$; likewise, $\sqrt[3]{2}$ is obtained as a solution of $x^3 - 2 = 0$.

we know as the ratio of the circumference to diameter of a circle, and e, the base of natural logarithms.

1.1.4 Decimal Numbers

Decimal numbers are so called because they use base 10 for counting.

Decimal numbers occur in:

* Measuring chemical properties, and interpreting chemical data.
* Defining relative atomic masses.
* Specifying the values of fundamental constants.

Decimal numbers consist of two parts separated by a **decimal point**:

* Digits to the left of the decimal point give the integral part of the number in units, tens, hundreds, thousands, *etc.*
* A series of digits to the right of the decimal point specify the fractional (or decimal) part of the number (tenths, hundredths, thousandths, *etc.*).

We can now more easily discuss the distinction between rational and irrational numbers, by considering how they are represented using decimal numbers.

Rational numbers, expressed in decimal form, may have either of the following representations:

* A finite number of digits after the decimal point. For example, $\frac{3}{8}$ becomes 0.375.
* A never-ending number of digits after the decimal point, but with a repeating pattern. For example, $\frac{70}{33}$ becomes 2.121 212 ..., with an infinite repeat pattern of "12".

Irrational numbers, expressed in decimal form have a never-ending number of decimal places in which there is *no* repeat pattern. For example, π is expressed as 3.141 592 653... and e as 2.718 281 82... As irrational numbers like π and e cannot be represented exactly by a finite number of digits, there will always be an error associated with their decimal representation, no matter how many decimal places we include. For example, the important irrational number e, which is the base for natural logarithms (not to be confused with the electron charge), appears widely in chemistry. This number is defined by the infinite sum of terms:

We can represent a sum of terms using a shorthand notation involving the summation symbol Σ. For example, the sum of terms $e = 1 + \frac{1}{1!} + \frac{1}{2!} + \frac{1}{3!} + \frac{1}{4!} + \cdots + \frac{1}{n!} + \cdots$ may be written as $\sum_{r=0}^{\infty} \frac{1}{r!}$ where the **counting index**, which we have arbitrarily named r, runs from 0 to ∞. A sum of terms which extends indefinitely is known as an **infinite series**, whilst one which extends to a finite number of terms is known as a **finite series**. Series are discussed in more detail in Chapter 1 of Volume 2.

$$e = 1 + \frac{1}{1!} + \frac{1}{2!} + \frac{1}{3!} + \frac{1}{4!} + \cdots + \frac{1}{n!} + \cdots \tag{1.1}$$

where $n!$ is the **factorial** (pronounced "n factorial") of the number n, defined as $n! = 1 \times 2 \times 3 \times 4 \cdots \times n$; for example, $4! = 1 \times 2 \times 3 \times 4$.

The form of equation (1.1) indicates that the value for e keeps becoming larger (but by increasingly smaller amounts) as we include progressively more and more terms in the sum, a feature clearly seen in Table 1.1, where the value for e has been truncated to 18 decimal places.

Table 1.1 An illustration of the effect of successive truncations to the estimated value of e derived from the infinite sum of terms given in equation (1.1)

n	Successive estimated values for e
1	2.000 000 000 000 000 000
5	2.716 666 666 666 666 666
10	2.718 281 801 146 384 797
15	2.718 281 828 458 994 464
20	2.718 281 828 459 045 235
25	2.718 281 828 459 045 235
30	2.718 281 828 459 045 235

Although the value of e has converged to 18 decimal places, it is still not exact; the addition of more terms causes the calculated value to change beyond the eighteenth decimal place. Likewise, attempts to calculate π are all based on the use of formulae with an infinite number of terms:

- Perhaps the most astonishing method uses only the number 2 and surds involving sums of 2:

$$\pi = 2 \times \frac{2}{\sqrt{2}} \times \frac{2}{\sqrt{2+2}} \times \frac{2}{\sqrt{2+2+2}} \times \cdots$$

- Another method involves an infinite sum of terms:

$$\frac{\pi}{2} = \frac{1}{1} + \frac{1 \times 1}{1 \times 3} + \frac{1 \times 1 \times 2}{1 \times 3 \times 5} + \frac{1 \times 1 \times 2 \times 3}{1 \times 3 \times 5 \times 7} + \cdots,$$

- A particularly elegant method uses a formula that relates the square of π to the sum of the inverses of the squares of all positive whole numbers:

$$\frac{\pi^2}{6} = 1 + \frac{1}{2^2} + \frac{1}{3^2} + \frac{1}{4^2} + \frac{1}{5^2} \cdots$$

However, this requires an enormous number of terms to achieve a satisfactory level of precision (see Chapter 1 in Volume 2 for more on infinite series and convergence).

Working with Decimal Numbers

As we have seen above, numbers in decimal form may have a finite, or infinite, number of digits after the decimal point. Thus, for example, we say that the number 1.4623 has four decimal places. However, since the decimal representations of irrational numbers, such as π or the surd $\sqrt{2}$, all have an infinite number of digits, it is necessary, when working with such decimal numbers, to reduce the number of digits to those that are **significant** (often indicated by the shorthand, "sig. figs."). In specifying the number of significant figures of a number displayed in decimal form, all zeros to the left of the first non-zero digit are taken as not significant and are therefore ignored. Thus, for example, both the numbers 0.1456 and 0.000 097 44 have four significant figures.

There are basically two approaches for reducing the number of digits to those deemed significant:

- **Truncation** of the decimal part of the number to an appropriate number of **decimal places** or significant digits. For example, we could truncate π, 3.141 592 653..., to seven significant figures (six decimal places) by dropping all digits after the 2, to yield 3.141 592. For future reference, we refer to the sequence of digits removed as the "tail" which, in this example, is 653...

- **Rounding up** or **rounding down** the decimal part of a number to a given number of decimal places is achieved by some generally accepted rules. The number is first truncated to the required number of decimal places, in the manner described above; attention is then focused on the tail (see above):

  (i) If the leading digit of the tail is greater than 5, then the last digit of the truncated decimal number is increased by unity (rounded up), *e.g.* rounding π to 6 d.p. yields 3.141 593.

  (ii) If the leading digit of the tail is less than 5, then the last digit of the truncated decimal number is left unchanged (the number is rounded down), *e.g.* rounding π to 5 d.p. yields 3.141 59.

  (iii) If the leading digit of the tail is 5, then:

    (a) If this is the only digit, or there are also trailing zeros, *e.g.* 3.7500, then the last digit of the truncated decimal number is rounded up if it is odd or down if it is even. Thus 3.75 is rounded up to 3.8 because the last digit of the truncated number is 7 and therefore odd, but 3.45 is rounded down to 3.4 because the last digit of the truncated number is 4 and therefore even. This somewhat complicated rule ensures that there is no bias in rounding up or down in cases where the leading digit of the tail is 5.

    (b) If any other non-zero digits appear in the tail, then the last digit of the truncated decimal number is rounded up, *e.g.* 3.751 is rounded up to 3.8.

Worked Problem 1.1

Q   Compare the results obtained by sequentially rounding 7.455 to an integer with the result obtained using a single act of rounding.

A   On applying the rules for rounding, the numbers produced in sequence are 7.46, 7.5, 8. Rounding directly from 7.455, we obtain 7.

Problem 1.1

Give the values of (a) 2.554 455, (b) 1.7232 0508, (c) π and (d) e to:
  (i) 5, 4 and 3 decimal places, by a single act of rounding in each case;
  (ii) 3 significant figures, using $\pi = 3.141\ 592\ 653$ and e = 2.718 281 828.

Observations on Rounding

Worked Problem 1.1 illustrates that different answers may be produced if the rules are not applied in the accepted way. In particular, sequential rounding is not acceptable, as potential errors may be introduced because more than one rounding is carried out. In general, it is accepted practice to present the result of a chemical calculation by rounding the result to the number of significant figures that are known to be reliable (zeros to the left of the first non-zero digit are not included). Thus, although π is given as 3.142 to four significant figures (three decimal places), $\pi/1000$ is given to four significant figures (and six decimal places) as 0.003142.

Rounding Errors

It should always be borne in mind that, in rounding a number up or down, we are introducing an error: the number thus represented is merely an approximation of the actual number. The conventions discussed above, for truncating and rounding a number, imply that a number obtained by rounding actually represents a range of numbers spanned by the implied error bound. Thus, π expressed to 4 decimal places, 3.1416, represents all numbers between 3.14155 and 3.14165, a feature that we can indicate by writing this rounded form of π as 3.14160 ± 0.00005. Whenever we use rounded numbers, it is prudent to aim to minimize the rounding error by expressing the number to a sufficient number of decimal places. However, we must also be aware that if we subsequently combine our number with other rounded numbers through addition, subtraction, multiplication or division, the errors associated with each number also combine, propagate and generally grow in size through the calculation.

Problem 1.2

(a) Specify whether each of the following numbers are rational or irrational and, where appropriate, give their values to four significant figures. You should assume that any repeat pattern will manifest itself within the given number of decimal places:
(i) 1.378 423 7842; (ii) 1.378 423 7842...; (iii) $\frac{1}{70}$; (iv) $\frac{\pi}{4}$; (v) 0.005068; (vi) $\frac{e}{10}$.

Note: the number e expressed to 9 decimal places, 2.718 281 828, appears to have a repeating pattern, which might wrongly suggest it is a rational number; however, if we extend to a further 2 decimal places, 2.718 281 828 46, we see that there is no repeating pattern and the number is irrational.

(b) In a titration experiment, the volume delivered by a burette is recorded as 23.3 cm^3. Give the number of significant figures, the number of decimal places and estimates for the maximum and minimum titres.

1.1.5 Combining Numbers

Numbers may be combined using the **arithmetic operations** of addition ($+$), subtraction ($-$), multiplication ($\times$) and division (/ or $\div$). The type of number (integer, rational, irrational) is not necessarily maintained under combination. Thus, for example, addition of the fractions 1/4 and 3/4 yields an integer, but division of 3 by 4 (both integers) yields the rational number (fraction) 3/4. When a number (say, 8) is multiplied by a fraction (say, 3/4), we say in words that we want the number which is three quarters *of* 8 which, in this case, is 6.

For addition and multiplication the order of operation is unimportant, regardless of the number of numbers being combined. Thus:

$$2 + 3 = 3 + 2$$

and

$$2 \times 3 = 3 \times 2$$

and we say both addition and multiplication are **commutative**. However, for subtraction and division, the order of operation *is* important, and we say that both are **non-commutative**:

$$2 - 3 \neq 3 - 2$$

and

$$\frac{2}{3} \neq \frac{3}{2}$$

One consequence of combining operations in an arithmetic expression is that ambiguity may arise in expressing the outcome. In such cases, it is imperative to include brackets (the generic term), where appropriate, to indicate which arithmetic operations should be evaluated first. The order in which arithmetic operations may be combined is described, by convention, by the **BODMAS** rules of precedence. These state that the order of preference is as follows:

Brackets
Of (multiplication by a fraction)
Division
Multiplication
Addition/Subtraction

For example:

- If we wish to evaluate $2 \times 3 + 5$, the result depends upon whether we perform the addition prior to multiplication or *vice versa*. The BODMAS rules tell us that multiplication takes precedence over addition and so the result should be $6 + 5 = 11$ and not $2 \times 8 = 16$. Using parentheses in this case removes any ambiguity, as we would then write the expression as $(2 \times 3) + 5$.
- If we wish to divide the sum of 15 and 21 by 3, then the expression $15 + 21/3$ yields the unintended result $15 + 7 = 22$, instead of 12, as division takes precedence over addition. Thus, in order to obtain the intended result, we introduce parentheses () to ensure that summation of 15 and 21 takes place before division:

$$(15 + 21)/3 = 36/3 = 12$$

Alternatively, this ambiguity is avoided by expressing the quotient in the form:

$$\frac{15 + 12}{3}$$

However, as the **solidus** sign, /, for division is in widespread use, it is important to be aware of possible ambiguity.

Powers or Indices

When a number is repeatedly multiplied by itself in an arithmetic expression, such as $3 \times 3 \times 3$, or $\frac{1}{2} \times \frac{1}{3} \times \frac{1}{4} \times \frac{1}{n}$, the **power** or **index** notation (also often called the **exponent**) is used to write such products in the forms 3^3 and $\left(\frac{3}{2}\right)^4$, respectively. Both numbers are in the general form a^n, where n is the index. If the index, n, is a positive integer, we define the number a^n as a raised to the nth power.

We can define a number of laws for combining numbers written in this form simply by inspecting expressions such as those given above: For example, we can rewrite the expression:

$$\frac{3}{2} \times \frac{3}{2} \times \frac{3}{2} \times \frac{3}{2} = \left(\frac{3}{2}\right)^4$$

as

$$\frac{3}{2} \times \frac{3}{2} \times \frac{3}{2} \times \frac{3}{2} = \left(\frac{3}{2}\right)^3 \left(\frac{3}{2}\right)^1 = \left(\frac{3}{2}\right)^4$$

and we see that the result is obtained simply by adding the indices of the numbers being combined. This rule is expressed in a general form as:

$$a^n a^m = a^{n+m} \tag{1.2}$$

For rational numbers, of the form $\frac{a}{b}$, raised to a power n, we can rewrite the number as a product of the numerator with a positive index and the denominator with a negative index:

$$\left(\frac{a}{b}\right)^n = \frac{a^n}{b^n} = a^n \times b^{-n} = a^n b^{-n} \tag{1.3}$$

which, in the case of the above example, yields:

$$\left(\frac{3}{2}\right)^4 = \frac{3^4}{2^4} = 3^4 \times 2^{-4}$$

On the other hand, if $b = a$, and their respective powers are different, then the rule gives:

$$\frac{a^n}{a^m} = a^n a^{-m} = a^{n-m} \tag{1.4}$$

The same rules apply for rational indices, as is seen in the following example:

$$\left(\frac{3}{2}\right)^{3/2} = \frac{3^{3/2}}{2^{3/2}} = 3^{3/2} \times 2^{-3/2}$$

Rational Powers

Numbers raised to powers $\frac{1}{2}, \frac{1}{3}, \frac{1}{4}, \ldots \frac{1}{n}$ define the square root, cube root, fourth root, ..., nth root, respectively. Numbers raised to the power m/n are interpreted either as the mth power of the nth root or as the nth root of the mth power. For example, $3^{m/n} = (3^{1/n})^m = (3^m)^{1/n}$. Numbers raised to a rational power may either simplify to an integer, for example $(27)^{1/3} = 3$, or may yield an irrational number, for example $(27)^{1/2} = 3 \times 3^{1/2} = 3^{3/2}$.

Further Properties of Indices

Consider the simplification of the expression $(3^2 \times 10^3)^2$:

$$\left(3^2 \times 10^3\right)^2 = 3^2 \times 10^3 \times 3^2 \times 10^3 = 3^4 \times 10^6$$

The above example illustrates the further property of indices that $(a^n)^m = a^{n \times m}$. Thus, we can summarize the rules for handling indices in equation (1.5):

$$a^n \times a^m = a^{n+m}; \ \frac{1}{a^n} = a^{-n}; \ \frac{a^n}{a^m} = a^{n-m}; \ (a^n)^m = a^{n \times m} = a^{nm}; \ a^0 = 1 \quad (1.5)$$

Note that, when multiplying symbols representing numbers, the multiplication sign ($\times$) may be dropped. For example, in the penultimate expression in equation (1.5), $a^{n \times m}$ becomes a^{nm}. In these kinds of expression, n and m can be integer or rational. Finally, if the product of two different numbers is raised to the power n, then the result is given by:

$$(ab)^n = a^n b^n \quad (1.6)$$

Problem 1.3

Simplify the following expressions:

(a) $\dfrac{10^2 \times 10^{-4}}{10^6}$; (b) $\dfrac{9 \times 2^4 \times 3^{-2}}{4^2}$; (c) $\left(\dfrac{10}{3^2 + 4^2 + 5^2}\right)^{-1/2}$; (d) $\dfrac{\left(2^4\right)^3}{4^4}$.

Worked Problem 1.2 and Problem 1.4 further illustrate how the (BODMAS) rules of precedence operate.

Worked Problem 1.2

Q Simplify the following expressions:

(a) $\frac{1}{2} \times (5-2) + 3 - 12/4 + 3 \times 2^2$.

(b) $\frac{1}{25} \times (3 \times 10^2)^2 + 3200/4 - 6 \times 120$.

A (a)

$$\frac{1}{2} \times (5-2) + 3 - \frac{12}{4} + 3 \times 2^2 = \frac{3}{2} + 3 - 3 + 12 = \frac{3}{2} + 12 = \frac{3}{2} + \frac{24}{2} = \frac{27}{2}$$

The penultimate step involves creating fractions with a common denominator, to make the addition easier.

(b)

$$\frac{1}{25}\left(3 \times 10^2\right)^2 + 3200/4 - 6 \times 120 = \frac{1}{25} \times (300 \times 300) + 800 - 720$$
$$= 3600 + 80 = 3680$$

In both parts, the rules of precedence are used to evaluate each bracket, and the results are combined using further applications of the rules.

Problem 1.4

Evaluate the following expressions, without using a calculator:

(a) $\left(2.5 \times 10^2 - 0.5 \times 10^2\right)^2 / 4 \times 10^4$; (b) $\left(\frac{1}{2 \times 4}\right)^{1/3} - 4 \times \frac{1}{16}$.

1.1.6 Scientific Notation

As has been noted earlier, many numbers occurring in chemical calculations are either extremely small or extremely large. Clearly, it becomes increasingly inconvenient to express such numbers using decimal notation, as the **order of magnitude** becomes increasingly large or small. For example, as seen in the introduction, the charge on an electron (in coulombs), expressed as a decimal number, is given by:

$$0.000\,000\,000\,000\,000\,000\,160\,2177$$

To get around this problem we can use **scientific notation** to write such numbers as a signed decimal number, usually with magnitude greater than or equal to 1 and less than 10, multiplied by an appropriate power of 10.

Thus, we write the fundamental unit of charge to nine significant figures as $1.602\,177\,33 \times 10^{-19}$ C.

Likewise, for very large numbers, like the speed of light, we write $c = 299\,792\,458$ m s^{-1}, which, in scientific notation, becomes $2.997\,924\,58 \times 10^8$ m s^{-1} (9 sig. figs.). Often we use $c = 3 \times 10^8$ m s^{-1}, using only one sig. fig., if we are carrying out a rough calculation.

Sometimes, an alternative notation is used for expressing a number in scientific form. Instead of specifying a power of 10 explicitly, it is common practice (particularly in computer programming) to give expressions for the speed of light and the charge on the electron as 2.998e8 m s^{-1} and 1.6e–19 C, respectively. In this notation, the number after the e is the power of 10 multiplying the decimal number prefix.

> The conventional representation of numbers using scientific notation is not always followed. For example, we might represent a bond length as 0.14×10^{-9} m rather than 1.4×10^{-10} m if we were referencing it to another measurement of length given in integer unit multiples of 10^{-9} m.

> The negative of the fundamental unit of charge is the charge carried by an electron.

Combining Numbers Given in Scientific Form

Consider the two numbers 4.2×10^{-8} and 3.5×10^{-6}; their product and quotient are given respectively by:

$$4.2 \times 10^{-8} \times 3.5 \times 10^{-6} = 14.7 \times 10^{-14} = 1.47 \times 10^{-13}$$

and

$$\frac{4.2 \times 10^{-8}}{3.5 \times 10^{-6}} = 1.2 \times 10^{-2}$$

However, in order to calculate the sum of the two numbers (by hand!), it may be necessary to adjust one of the powers of ten, to ensure equality of powers of 10 in the two numbers. Thus, for example:

$$4.2 \times 10^{-8} + 3.5 \times 10^{-6} = 0.042 \times 10^{-6} + 3.5 \times 10^{-6} = 3.542 \times 10^{-6}$$

Names and Abbreviations for Powers of Ten

As we have seen in some of the examples described above, an added complication in performing chemical calculations often involves the presence of units. More often than not, these numbers may be expressed in scientific form, and so, in order to rationalize and simplify their specification, it is conventional to use the names and abbreviations given in Table 1.2, adjusting the decimal number given as prefix as appropriate.

Table 1.2 Names and abbreviations used to specify the order of magnitude of numbers expressed in scientific notation

10^{15}	10^{12}	10^9	10^6	10^3	10^{-1}	10^{-2}	10^{-3}	10^{-6}	10^{-9}	10^{-12}	10^{-15}	10^{-18}
peta	tera	giga	mega	kilo	deci	centi	milli	micro	nano	pico	femto	atto
P	T	G	M	k	d	c	m	μ	n	p	f	a

Thus, for example:

- The charge on the electron is given as 0.16 aC, to two significant figures.
- The binding energy of the electron in the hydrogen atom is given by 2.179×10^{-18} J, which is specified as 2.179 aJ.
- The bond vibration frequency for HF is 1.2404×10^{14} s^{-1}, which is given as 124.04 Ts^{-1} or 0.124 04 Ps^{-1}.

Some of these data are used in Problem 1.5.

Problem 1.5

(a) Given that 1 eV of energy is equivalent to 0.1602 aJ, use the information given above to calculate the ionization energy of the hydrogen atom to 3 sig. figs. in electronvolts (a common macroscopic energy unit).

(b) Given that the Planck constant, h, has the value 6.626×10^{-34} J s, and that vibrational energy, ε_{vib}, is related to vibration frequency, v, according to $\varepsilon_{vib} = hv$, calculate the value of ε_{vib} for HF to 3 sig. figs.

The results you obtain for Problem 1.5 should show that, since the joule (J) is a macroscopic base unit of energy, property values on the microscopic scale have extremely small magnitudes. We now explore this idea further in Worked Problem 1.3 and in Problem 1.6 to give more practice in manipulating numbers in scientific form and, more importantly, to provide further insight into size differences in the microscopic and macroscopic worlds.

Worked Problem 1.3

Q   Calculate the whole number of football pitches that would be covered by 1 mol of benzene, spread out in a molecular monolayer, assuming that: (a) a benzene molecule can be considered as a disc of radius $r = 300$ pm and (b) the area of a soccer pitch is 6900 m^2.

A   The number of molecules in 1 mol of benzene is equal to 6.022×10^{23} (to four sig. figs.). If we assume that a reasonable estimate for the area, A, covered can be calculated by multiplying the effective area of each molecule, a, by the number of molecules in 1 mol, then:

$$A = 6.022 \times 10^{23} \times \pi \times (300 \times 10^{-12}\,m)^2 = 1.70 \times 10^5\,m^2$$

where $a = \pi \times r^2$ for the area of a disc of radius r. Thus, if we now divide A by the area of a football pitch, we obtain an estimate for the area covered in terms of the number of football pitches, n, covered by 1 mol of benzene:

$$n = \frac{A}{FP} = \frac{1.70 \times 10^5}{6900} = 24.6$$

where the units m^2 in numerator and denominator cancel. There is evidently enough benzene to cover 25 football pitches.

Problem 1.6

Given that the density, ρ, of benzene is 879 kg m^{-3} at 298 K, show that 1 cm^3 of benzene contains 0.0113 mol (assume a molar mass for benzene of 0.078 kg mol^{-1}).

Problem 1.7

Assuming that the Earth's radius is 6378 km, and that a molecule of benzene may be treated as a disc of radius 300 pm, calculate the mass of benzene needed to create a chain of molecules around the equator of the Earth.

Worked Problem 1.4

Q Gold crystallizes in a cubic close-packed structure, based on a cube with side a. Given that the density of metallic gold is 19.3×10^3 kg m^{-3} (density is mass divided by volume) and that $a = 4.08 \times 10^{-10}$ m, find the number of Au atoms per unit cell using the molar mass of ^{197}Au(100%) = 197 g mol^{-1}. Give your answer to 3 sig. figs.

A The mass, m, of the unit cell is given by volume times density; thus:

$$m = (4.08 \times 10^{-10}m)^3 \times 19.3 \times 10^3\,kg\,m^{-3} = 1.31 \times 10^{-24}\,kg$$
$$= 1.31 \times 10^{-21}\,g.$$

If $M(\text{Au}) = 197$ g mol^{-1}, then the mass of one gold atom is:

$$\frac{197 \, \text{g mol}^{-1}}{6.02 \times 10^{23} \, \text{mol}^{-1}} = 3.27 \times 10^{-22} \, \text{g}$$

Thus, the number of atoms per unit cell is:

$$n = \frac{1.31 \times 10^{-21} \, \text{g}}{3.27 \times 10^{-22} \, \text{g}} = 4.01$$

Cubic close packing implies 4 atoms per unit cell, so this result would seem to conform to our expectation in this case. The deviation from the theoretically correct result of 4 derives simply from the precision with which we specify the density, the value for a, the molar mass of Au and the Avogadro number. If we had chosen to specify some or all of these quantities to 4 or 5 significant figures rather than 3, then we might reasonably expect to achieve a better match. It is worth remarking that in carrying out this type of calculation we need also consider whether the metal in question is isotopically pure or not. For example, metallic copper, which also crystallizes in a cubic close-packed structure, is made up of a mixture of ^{63}Cu and ^{65}Cu. In this case, the calculated number of Cu atoms per unit cell will be affected by the difference in isotopic masses in the sample because we assume that each unit cell has the same isotopic composition (since the definition of a unit cell describes it as the basic repeating unit).

Problem 1.8

Calculate the number of Au atoms per unit cell to 3 d.p., given that the molar mass of $^{197}\text{Au}(100\%) = 196.97$ g mol^{-1}, the density of metallic gold is 19.321×10^3 kg m^{-3}, that $a = 4.0783 \times 10^{-10}$ m and given Avogadro's constant $= 6.0221 \times 10^{23}$ mol^{-1}.

The association of numbers with patterns was common in ancient Greece. The so-called figurate numbers 3, 6, 10 and 15 given in the text are examples of triangular numbers; other examples include square (1, 4, 9, 16, 25,...) and pentagonal (1, 5, 12, 22, 35,...) numbers.

1.1.7 Relationships between Numbers

Frequently in chemistry we find ourselves considering the significance of a numerical quantity, associated with some property of a system, in terms of its relationship to some accepted standard. For example, we might measure a rate constant which tells us whether a particular reaction is fast or slow. We can only draw a conclusion in this respect by comparing it to some standard which we know to imply one extreme or the other or somewhere in between. Of course, this activity is important in all areas of life, and highlights the value of being able to assess how numbers relate to one another. Historically, this relationship has been made easier by associating numbers with patterns (see Figure 1.1).

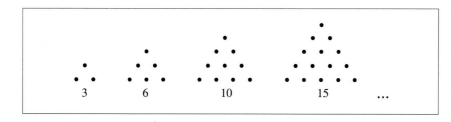

3   6   10   15   ...

Figure 1.1 The relationship between numbers is made easier by associating them with patterns. In general, numbers which can be represented in this way are called figurate numbers. In this example, the numbers 3, 6, 10 and 15 are known as triangular numbers and their relative magnitude is easily seen by the increasing number of dots used to represent them

These so-called figurate numbers (in this case, triangular numbers) are more easily presented in order of increasing magnitude, simply because it is easy to see that there are more blobs to the right than to the left. By following this convention, it is then straightforward to deduce the next number in the sequence (here 21, by constructing a triangle with a row of six blobs at the base. Intuitively, we can see that 6 is of greater ($>$) magnitude than 3, and of lesser ($<$) magnitude than 15, simply by counting blobs. The mathematical notation for describing these two relations is $6 > 3$ and $6 < 15$. Such relations are termed **inequalities**. Note that it is equally true that $3 < 6$ and $15 > 6$. We can also combine these two relations into one: either $3 < 6 < 15$ or $15 > 6 > 3$.

Problem 1.9

Use the inequality symbols ($<$, $>$) to express two relationships between the following pairs of numbers: (a) 2, 6; (b) 1.467, 1.469; (c) π, e.

Negative Numbers

The question of negative numbers must now be addressed. All negative numbers are less than zero, and hence we can say immediately that $-6 < 3$. Furthermore, as $6 > -3$, we can obtain the latter inequality from the former simply by changing the sign of the two numbers (multiplying through by -1) *and* reversing the inequality sign.

Problem 1.10

(a) Use inequalities to express the fact that a number given as 2.456 lies between 2.4555 and 2.4565.
(b) Express each of the following inequalities in alternative ways: $-5.35 < -5.34$; $-5.35 < 5.34$.

Very Large and Very Small Numbers

The numerical value of the Avogadro constant is 6.022×10^{23}, a very large number. An expression of the disparity in the size between this number and unity may be expressed in the form $6.022 \times 10^{23} \gg 1$; likewise, for the magnitude of the charge on the electron, we can express its smallness with respect to unity as $1.602 \times 10^{-19} \ll 1$.

Infinity

The concept of an unquantifiably enormous number is of considerable importance to us in many contexts, but probably is most familiar to us when we think about the size of the universe or the concept of time as never ending. For example, the sums of the first 100, 1000 and 1000 000 positive integers are 5050, 500 500, and 500 000 500 000, respectively. If the upper limit is extended to 1000 000 000, and so on, we see that the total sum increases without limit. Such summations of numbers – be they integers, rationals, or decimal numbers – which display this behaviour are said to tend to **infinity**. The use of the symbol ∞ to designate infinity should not be taken to suggest that infinity is a number: it is not! The symbol ∞ simply represents the concept of indefinable, unending enormity. It also arises in situations where a constant is divided by an increasing small number. Thus, the sequence of values $\frac{1}{10^{-6}}, \frac{1}{10^{-20}}, \frac{1}{10^{-1000}}, \ldots$ (that is, 10^6, 10^{20}, 10^{1000}) clearly tends to infinity, whilst the same sequence of negative terms tends to $-\infty$. Once again, there is no limiting value for the growing negative number $-\frac{1}{10^{-n}}$ as the value of n increases (the denominator decreases towards zero). Although it is tempting to write $\frac{1}{0} = \infty$, this statement is devoid of mathematical meaning because we could then just as easily write $\frac{2}{0} = \infty$, which would imply that $1 = 2$, which is clearly not the case. We shall see in Chapter 3 how to evaluate limiting values of expressions in which the denominator approaches zero.

The Magnitude

The magnitude of a number is always positive, and is obtained by removing any sign. Thus, the magnitude of -4.2 is given using the **modulus** notation as $|-4.2| = 4.2$.

Problem 1.11

(a) Give the values of $|4 - 9|$, $|-3 - 6|$, $|9 - 4|$.

(b) Give the limiting values of the numbers 10^{-m}, 10^{m}, -10^{-m}, -10^{m}, as m tends to infinity.

1.2 Algebra

Much of the preceding discussion has concerned numbers and some of the laws of **arithmetic** used for their manipulation. In practice, however, we do not generally undertake arithmetic operations on numbers obtained from some experimental measurement at the outset: we need a set of instructions telling us how to process the number(s) to obtain some useful property of the system. This set of instructions takes the form of a **formula** involving **constants**, of fixed value, and **variables** represented by a symbol or letter: the symbols designate quantities that, at some future stage, we might give specific numerical values determined by measurements on the system. Formulae of all kinds are important, and their construction and use are based on the rules of **algebra**. The quantity associated with the symbol is usually called a variable because it can take its value from some given set of values. These variables may be **continuous variables** if they can take any value from within some interval of numbers (for example, temperature or concentration), or they may be **discrete variables** if their value is restricted to a discrete set of values, such as a subset of positive integers (for example, atomic number). One further complicating issue is that, in processing a number associated with some physical property of the system, we also have to consider the units associated with that property. In practice, the units are also processed by the formula, but some care is needed in how to present units within a formula (an issue discussed later on in Chapter 2). However, the most important point is that algebra provides us with a tool for advancing from single one-off calculations to a general formula which provides us with the means to understand the chemistry. Without formulae, mathematics and theory, we are in the dark!

Symbols used to represent variables or constants in a formula may additionally have subscripts and/or superscripts attached. For example, x_1, k_2, A^{12}.

1.2.1 Generating a Formula for the Sum of the First n Positive Integers

Consider first the simple problem of summing the integers 1, 2, 3, 4 and 5. The result by arithmetic (mental or otherwise) is 15. However, what if we want to sum the integers from 1 to 20? We can accomplish this easily enough by typing the numbers into a calculator or adding them in our head, to obtain the result 210, but the process becomes somewhat more tedious. Now, if we want to sum the sequence of integers from 1 to some, as yet unspecified, upper limit, denoted by the letter n, we need a formula that allows us to evaluate this sum without actually having to add each of the numbers individually. We can accomplish this as follows:

- Write down the sum of the first five integers, 1 to 5, from highest to lowest, and introduce the symbol S_5 to represent this sum:

$$S_5 = 5 + 4 + 3 + 2 + 1$$

- Repeat the exercise by summing the same five integers from lowest to highest:

$$S_5 = 1 + 2 + 3 + 4 + 5$$

- Add the two expressions to obtain:

$$2S_5 = 6 + 6 + 6 + 6 + 6 = 5 \times 6 = 30 \Rightarrow S_5 = 15$$

where the symbol $\Rightarrow$ means "implies".

If we repeat this procedure for the first six integers, rather than the first five, we obtain:

$$2S_6 = 7 + 7 + 7 + 7 + 7 + 7 = 6 \times 7 = 42 \rightarrow S_6 = 21$$

We can see that, in each case, the respective sum is obtained by multiplying the number of integers, n, in the sum by the same number incremented by 1, and dividing the result by 2; that is:

$$S_5 = \frac{5 \times 6}{2} = 15$$

$$S_6 = \frac{6 \times 7}{2} = 21$$

Problem 1.12

Use the result given above to deduce expressions that yield the sum of the first (a) 100 positive integers and (b) 68 negative integers.

The pattern should now be apparent, and we can generalize the expression for the sum of the first n positive integers by multiplying n by $n+1$, and dividing the result by 2:

$$S_n = \frac{n \times (n+1)}{2}$$

It is usual practice, when symbols are involved, to drop explicit use of the multiplication sign $\times$, thus enabling the formula for S_n to be given in the form:

$$S_n = \frac{n(n+1)}{2} \tag{1.7}$$

We can test our new formula, by using it to determine the sum of the positive integers from 1 to 20:

$$S = \frac{20 \times 21}{2} = 210$$

1.2.2 Algebraic Manipulation

The rules for manipulating algebraic symbols are the same as those for numbers. Thus we can formally add, subtract, multiply and divide combinations of symbols, just as if they were numbers. In the example given above, we have used parentheses to avoid ambiguity in how to evaluate the sum. The general rules for expanding expressions in parentheses ( ), brackets [ ] or braces { } take the following forms:

$$a(b+c) = ab + ac = (b+c)a \tag{1.8}$$

and

$$(b+c)/d = b/d + c/d \tag{1.9}$$

When we want to multiply two expressions in parentheses together, the first rule is simply applied twice. Thus, if we are given the expression

$$(a+b)(c+d)$$

we can expand by letting

$$X = (a+b)$$

and then:

$$(a+b)(c+d) = X(c+d) = Xc + Xd = (a+b)c + (a+b)d$$
$$= ac + bc + ad + bd \tag{1.10}$$

We can use these rules to expand our expression for the sum of n integers above to obtain either:

$$S_n = n^2/2 + n/2 \text{ or } S_n = n/2 + n^2/2$$

However, it would be usual in this case to stick to our original expression because it is more compact and aesthetically pleasing.

The ordering of symbols representing numerical quantities in product and summation forms is of no consequence: the symbols **commute** under both addition and multiplication.

Worked Problem 1.5

A formula of chemical importance

Q   The spin of a proton can have two orientations with respect to the direction of a homogeneous magnetic field, either "spin up" or "spin down", often represented by the arrows ↑ or ↓, respectively. In an NMR experiment, the two orientations have different energies. Let us now consider how many possible spin combinations there are for two and for three equivalent protons, and then derive the result for n equivalent protons.

A   For two equivalent protons, the first proton can have one of two possible spin states, spin up or spin down. Each of these can combine with either one of the spin states possible for the second proton. Thus the total number of two-spin state combinations is 2×2^2 (a result which is more useful than the arithmetic result, 4). The four two-spin states are as follows:

First spin stateup:	↑↑	↑↓
First spin statedown:	↓↑	↓↓

If we now include a third equivalent proton, as would be appropriate for the methyl group, each parent two-spin state gives rise to 2 three-spin states, thus doubling the overall number of spin states to yield a total of $2 \times 2^2 = 2^3$, which is, of course, 8:

First two-spin state:	↑↑↑	↑↑↓
Second two-spin state:	↑↓↑	↑↓↓
Third two-spin state:	↓↑↑	↓↑↓
Fourth two-spin state:	↓↓↑	↓↓↓

It should now be apparent that, for each additional equivalent proton, the number of spin states is doubled: thus, for 4 protons, there are 2^4 spin states and, hence, for n equivalent protons, there are 2^n spin states.

Problem 1.13

The nuclear spin of the deuteron, which has a spin quantum number I of 1, can have three orientations, represented symbolically by $\uparrow, \rightarrow, \downarrow$. The expression derived above for two-spin state systems can readily be extended to any number of spin states. Thus for n equivalent nuclei having m possible spin orientations, the number of possible spin states is given by m^n.

(a) Deduce the number of three-spin states for the CD_2 radical (where D is the chemical symbol for deuterium). Now do the same for a fully deuterated methyl radical.

(b) Given that there are $2I + 1$ spin states for a nuclear spin quantum number, I, give the number of spin states associated with n equivalent nuclei with spin I.

(c) State the number of nuclear spin states for a single atom of ^{51}V, which has a nuclear spin quantum number of $\frac{7}{2}$ (this is useful for understanding the electronic characteristics of complexes of vanadium).

The deuteron is the name given to the nucleus of deuterium which contains one proton and one neutron.

Dealing with Negative and Positive Signs

In the algebraic expressions considered so far, all the constituent terms carried a positive sign. In general, however, we have to work with expressions involving terms carrying positive or negative signs. Dealing with signed terms is straightforward when we appreciate that a negative or positive sign associated with a number or symbol simply implies the **operation** "multiply by -1 or $+1$", respectively. For example, the operation:

$$(-a) \times (-b)$$

is equivalent to writing:

$$-1 \times a \times -1 \times b = (-1 \times -1) \times a \times b = ab$$

A simple set of rules can be constructed, using this reasoning, to help us to carry out multiplication and division of signed numbers or symbols:

Multiplication	*Division*
$[(+a) \times (+b)] = ab$	$[(+a)/(+b)] = a/b$
$[(+a) \times (-b)] = -ab$	$[(+a)/(-b)] = -a/b$
$[(-a) \times (+b)] = -ab$	$[(-a)/(+b)] = -a/b$
$[(-a) \times (-b)] = ab$	$[(-a)/(-b)] = a/b$

These rules are valid if a and b are numbers, symbols or algebraic expressions.

Worked Problem 1.6

Q   Given that $x = 6$, $y = -3$, $z = 2$, find the value of each of the following algebraic expressions:

  (a) $xy + 2y - 3z/y$; (b) $(x + 2z)(z - y)$.

A   (a) $xy + 2y - 3z/y = -18 - 6 + 6/3 = -22$; (b) $(x + 2z)(z - y) = 10 \times 5 = 50$.

Problem 1.14

Given that $u = (x + y)$, $v = (x - y)$, find an expression for:
  (a) $(u^2 + v^2)/(v - u)$; (b) $uv/(2u - v)$; (c) $10^{u+v}/10^{u-v}$.

Problem 1.15

Simplify the following expressions:
  (a) (i) $4p - q - (2q + 3p)$; (ii) $3p^2 - p(4p - 7)$;
  (b) (i) $(1 + x)^2 - (1 - x)^2$; (ii) $x(2x + 1) - (1 + x - x^2)$.

Working with Rational Expressions

A rational expression (often called a **quotient**) takes the form $\frac{a}{b}$, where a and b may be simple or complicated expressions. In many instances it is necessary to simplify the appearance of such expressions by searching for **common factors** (symbols or numbers common to each term) and, if necessary, by deleting such factors in both the numerator and denominator. For example, in

$$\frac{3x^2 - 12xy}{3}$$

the numerator has 3 and x as common factors, whilst the denominator has 3. Since the denominator and numerator both have the common factor 3, this may be cancelled from each term to give:

$$\frac{x^2 - 4xy}{1} = x^2 - 4xy$$

which simplifies further to $x(x - 4y)$, once the common factor x has been removed from each term. In this case, the rational expression reduces to a simple expression. We should also be aware that, whenever we are faced with a rational expression involving symbols, it is necessary to specify that any symbol appearing as a cancellable common factor, in both numerator and denominator, cannot take the value zero, because otherwise the resulting expression would become $\frac{0}{0}$, which is indeterminate (*i.e.* meaningless!).

Problem 1.16

Simplify the following expressions, and indicate any restrictions on the symbol values:

(a) $\dfrac{p^4 q^2}{p^2 q^3}$; (b) $\dfrac{p^8 q^{-3}}{p^{-5} q^2}$; (c) $\dfrac{4x}{6x^2 - 2x}$; (d) $\dfrac{3x^2 - 12xy}{3}$.

1.2.3 Polynomials

A **polynomial** is represented by a sum of symbols raised to different powers, each with a different coefficient. For example, $3x^3 - 2x + 1$ involves a sum of x raised to the third, first and zeroth powers (remember that $x^0 = 1$) with coefficients 3, -2, and 1, respectively. The highest power indicates the **degree** of the polynomial and so, for this example, the expression is a polynomial of the third degree.

Factorizing a Polynomial

Since x does not appear in all three terms in the polynomial $3x^3 - 2x + 1$, it cannot be a common factor; however, if we can find a number a, such that $3a^3 - 2a + 1 = 0$, then $x - a$ *is* a common factor of the polynomial. Thus in order to factorize the example given, we need first to solve the expression:

$$3a^3 - 2a + 1 = 0$$

Trial and error shows that $a = -1$ is a solution of this equation, which means that $x - (-1) = x + 1$ is a factor of $3x^3 - 2x + 1$. It is now possible to express $3x^3 - 2x + 1$ in the form $(x + 1)(3x^2 - 3x - 1)$. Note that the second degree polynomial in parentheses does in fact factorize further, but the resulting expression is not very simple in appearance.

Problem 1.17

Express the following polynomials in factored form, indicating any limitations on the values of x:

(a) (i) $x^2 - 3x + 2$; (ii) $x^3 - 7x + 6$; (b) (i) $\dfrac{x^3 - 7x + 6}{x - 2}$; (ii) $\dfrac{x^2 - 1}{x - 1}$.

Forming a Common Denominator

An expression of the form $\frac{x}{a} + \frac{y}{b}$ may be written as one rational expression with a common **denominator** ab as follows:

$$\frac{x}{a} + \frac{y}{b} = \frac{xb + ya}{ab} \qquad (1.11)$$

If there are three terms to combine, we reduce the first two terms to a rational expression, and then repeat the process with the new and the third terms.

Problem 1.18

Express the following in common denominator form, in which there are no common factors in numerator and denominator:

(a) (i) $\frac{3x}{4} - \frac{x}{2}$; (ii) $\frac{2}{x} - \frac{1}{x^2}$; (iii) $1 - \frac{1}{x} + \frac{2}{x^2}$;

(b) (i) $\frac{1}{1+x} - \frac{1}{1-x}$; (ii) $\frac{2x}{x^2+1} - \frac{2}{x}$.

1.2.4 Coping with Units

In chemistry, we work with algebraic expressions involving symbols representing particular properties or quantities, such as temperature, concentration, wavelength and so on. Any physical quantity is described in terms not only of its **magnitude** but also of its **dimensions**, the latter giving rise to units, the natures of which are determined by the chosen system of units. In chemistry, we use the SI system of units. For example, if we specify a temperature of 273 K, then the dimension is temperature, usually given the symbol T, the magnitude is 273 and the base **unit** of temperature is the Kelvin, with name K. Similarly, a distance between nuclei of 150 pm in a molecule has dimensions of length, given the symbol l, a magnitude of 150 and a unit of pm (10^{-12} m). All such physical quantities must be thought of as the product of the magnitude, given by a number, and the appropriate unit(s), specified by one or more names. Since each symbol representing a physical quantity is understood to involve a number and appropriate units (unless we are dealing with a pure number like percentage absorbance in spectroscopy), we treat the property symbols and unit names as algebraic quantities. All the usual rules apply and, for example, in the case of:

- The molar energy property, E, we may wish to use the rules of indices to write $E = 200$ kJ/mol as 200 kJ mol^{-1}.
- Concentration, c, we are concerned with amount of substance (name n, unit mol) per unit volume (name V, unit m^3).

If necessary, as seen earlier, we can manipulate the various prefixes for the SI base units as required. Further practice is given in the following problem.

Problem 1.19

(a) The expression RT/F occurs widely in electrochemistry. The gas constant, R, has units J K^{-1} mol^{-1}, the Faraday constant, F, has units C mol^{-1}, and the temperature, T, is measured in K. Given that 1 coulomb volt (C V) is equivalent to 1 joule (J), find the units of the given expression.

(b) The Rydberg constant, $R_\infty = \dfrac{m_e e^4}{8h^3 c \varepsilon_0^2}$, occurs in models used for interpreting atomic spectral data; m_e is the mass of the electron (kg), e is the elementary charge (C), c the speed of light (m s^{-1}), ε_0 the vacuum permittivity (J^{-1} C^2 m^{-1}) and h the Planck constant (J s). Given that 1 J is equivalent to 1 kg m^2 s^{-2}, find the units of R_∞.

Summary of Key Points

This chapter has revisited the elementary but important mathematical concepts of numbers and algebra as a foundation to the following chapter on functions and equations. The key points discussed include:

1. The different types of number: integers, rational, irrational and decimal.

2. The rules for rounding decimal numbers.

3. The rules for combining numbers, powers and indices.

4. Scientific notation for very large and very small numbers.

5. The relationships between numbers: how we reference numbers with respect to magnitude and sign.

6. The principles of algebra, generating a formula and algebraic manipulation.

7. Working with polynomials.

8. An introduction to units.

Reference

1. R. N. Perutz, *Univ. Chem. Educ.*, 1999, **3**, 37.

Further Reading

D. Wells, *The Penguin Dictionary of Curious and Interesting Numbers*, Penguin, London, 1987.

2

Functions and Equations: Their Form and Use

As we saw in Chapter 1, the importance of numbers in chemistry derives from the fact that experimental measurement of a particular chemical or physical property will always yield a numerical value to which we attach some significance. This might involve direct measurement of an intrinsic property of an atom or molecule, such as ionization energy or conductivity, but, more frequently, we find it necessary to use theory to relate the measured property to other properties of the system. For example, the rotational constant, B, for the diatomic molecule CO can be obtained directly from a measurement of the separation of adjacent rotational lines in the infrared spectrum. Theory provides the link between the measured rotational constant and the moment of inertia, I, of the molecule by the formula:

$$B = \frac{h}{8\pi^2 I c}$$

where h is Planck's constant and c is the speed of light. The moment of inertia itself is related to the square of the bond length of the molecule by:

$$I = \mu r^2$$

where μ is the reduced mass. The relationship between B and r was originally derived, in part, from the application of quantum mechanics to the problem of the rigid rotor. In general, relationships between one chemical or physical property of a system and another are described by mathematical functions. Such functions are especially important for building the mathematical models we need to predict changes in given property values that result from changes in the parameters defining the system. If we can predict such changes, then we are well on the way to understanding our system better! However, before we can explore these applications further, we have to define function in its mathematical sense. This is a necessary step because, in chemistry, the all-pervasive presence of units complicates the issue.

For a diatomic molecule AB the reduced mass is given by $\mu = \frac{m_A m_B}{m_A + m_B}$, where m_A and m_B are the masses of A and B, respectively.

Aims

This chapter aims to demonstrate the importance of mathematical functions and equations in a chemical context. By the end of the chapter you should be able to:

- Work with functions in the form of a table, formula or prescription and, for each type, specify the independent and dependent variables, an appropriate domain, and construct a suitable graphical plot using Cartesian coordinates
- Recognize periodic, symmetric or antisymmetric character in a function
- Find the factors and roots of simple polynomial equations using either algebraic or graphical procedures
- Use the laws of algebra to simplify expressions of all kinds
- Work with units using algebra
- Use the formulae for the logarithm of a product or quotient of expressions
- Understand the difference between degrees and radians for measurement of angles
- Be familiar with the use of trigonometric identities and addition formulae

2.1   Defining Functions

Our aim in this section is to show what features need to be understood in order to define a function properly as a mathematical object. First of all, let us consider the association between an arbitrary *number*, x, and the *number* $2x + 1$. We can thus associate the number 6 with 13, π with $2\pi + 1$, 1.414 with 3.828, and so on. It is conventional practice to express this association as a formula, or equation:

$$y = 2x + 1$$

where the unspecified number, x, is the input number for the formula, and y the output number. Before we can say that this association expresses y as a **function** of x, we need to:

- Specify the set of numbers for which the formula applies (the **domain**).
- Check that each value of x is associated with only *one* value of y.

A **subset** of R or I consists of a selection of real numbers or integers, respectively. Of course, I itself is by definition a subset of R.

In the present example, we could specify the domain as either the collection of all real numbers (conventionally described as the **set R**), or the set of all integers, **I**, or a subset of either or both, thereby satisfying the first requirement.

We can also see in this case that any number chosen as input generates a single number, y, as output and so the second requirement is also satisfied. It is very important to keep in mind that the function is defined not only by the formula but also by the domain; consequently, if we specify that the formula $y = 2x + 1$ applies to any real number as input, x, then we can define the function $y = f(x)$ where $f(x) = 2x + 1$ with domain **R**. In this case, f is the name of the function that describes both the formula *and* the permitted values of x. If we had specified that the same formula $y = 2x + 1$ is used with the domain consisting of all *integers*, **I**, then we would be dealing with a *different* function, which we might wish to call $y = g(x)$. For both functions, y is the number produced by the formula for a given x defined within the specified domain of the function. The association between x and y defines different functions for different subsets of numbers, even though the formula of association is the same. Where a function is defined by a formula, and the domain is not explicitly stated, then it is assumed that the domain consists of all real numbers for which the function has a real value. This is called the **natural domain** of the function.

Worked Problem 2.1

Q Define two possible domains of the function $f(x) = \dfrac{1}{(x-4)(x+3)}$.

A The value of $f(x)$ is indeterminate for $x = 4$ and $x = -3$ because division by zero yields an indeterminate result. For all other values of x the function is defined, and consequently the domain of $f(x)$ could be either all real numbers, excluding $x = 4$ and $x = -3$, or all integers excluding $x = 4$ and $x = -3$. We could write this explicitly as either:

$$f(x) = \frac{1}{(x - 4)(x + 3)}, x \in \mathbf{R},\, x \neq 4,\, x \neq -3$$

or:

$$f(x) = \frac{1}{(x - 4)(x + 3)}, x \in \mathbf{I},\, x \neq 4,\, x \neq -3$$

It is often unnecessary to specify the domain explicitly, as the restrictions are evident from the formula.

The symbol $\in$ means "is an element of" or alternatively "belongs to". Thus $x \in \mathbf{R}$ means that "x is an element of the set of real numbers".

The symbols x and y, used in the function formulae, are conventionally termed the **independent variable** and **dependent variable**, respectively. This terminology conveys the idea that we are free to assign values to the independent variable but that, once we have done so, a unique value for the dependent variable results. A function may have more than one independent variable, in which case a domain needs to be specified for each variable. For example, the formula:

$$y = pq/r$$

expresses an association between the three independent variables p, q and r and the dependent variable y. The domain is defined by specifying the permitted values associated with each of the independent variables. Having checked that only one value of the dependent variable results for a given set of values of p, q and r, we may then define the function:

$$y = f(p, q, r) = pq/r \qquad (2.1)$$

In practice, although many functions that we meet in a chemical context have more than one independent variable, the function may be reduced to a single variable by specifying that one or more of the other variables remain constant.

Frequently, we specify functions by formulae that do not explicitly involve a dependent variable but express the function simply in terms of the formula and the label used to denote the function. For example, the formula $f(x) = 2x + 1$ defines the function f that associates the number $2x + 1$ with the number x. Thus, $f(-5) = -9$ implies that f associates -9 with -5, while $f(3) = 7$ implies that f associates 7 with 3. Although this way of presenting the function does not involve a dependent variable, we can introduce one by letting $y = f(x)$, and rewriting the function as $y = 2x + 1$. The most frequently encountered labels used for the independent and dependent variables are x and y, respectively, but these labels are entirely arbitrary. We might just as easily use the labels p and r or ϕ and ρ; similarly, when labelling the function, instead of f we might use g, h, F or ψ, or indeed any label which we think appropriate. For example, if we wanted to collectively label the group of 1s, 2s, 3s, ... atomic orbital functions, we might use the name ϕ, and then distinguish each function using a numbered suffix, say ϕ_1, ϕ_2, ϕ_3. Our choice here is entirely arbitrary, but is designed to allow us, in this case, to group similar types of functions under a common name. You could rightly argue that the labels ϕ_1, ϕ_2, ϕ_3 provide little that the labels 1s, 2s, 3s, ... do not. The point here is that either will do, and it is really just a matter of taste, context, convenience or convention that dictates what labels and names we use. It is very important that we do not allow unfamiliar labels to give the impression that an otherwise straightforward association or function is more complicated than it actually is!

Worked Problem 2.2

Q The energies of the electron orbits in Bohr's model of the hydrogen atom take discrete values according to the expression

$$E_n = \frac{-m_e e^4}{8\varepsilon_0^2 h^2} \frac{1}{n^2}$$

where m_e and e are the mass and charge of the electron, respectively, h is Planck's constant and ε_0 the vacuum permittivity. Write down an equivalent expression using the labels x and y for the independent and dependent variables, respectively, and any labels you feel appropriate for the constants.

A The formula has exactly the same mathematical form as the formula $y = -ax^{-2}$. The energy E_n is the dependent variable, equivalent to y in the second formula and carries units of J. The principal quantum number, n, is the independent variable, equivalent to x, and can only take integer values greater than or equal to 1. The constants m_e, e, h and ε_0 can be collected together into a single constant, equivalent to a in the second formula. The domain of the original formula is restricted to all positive integers. However, this restriction may not necessarily apply to the second formula, which might have as domain all real numbers, for example. Consequently, although the formulae are identical in form, we define two distinct functions, which are distinguished from each other by their domains.

The requirement that a function be single valued, for a given input value for the independent variable, will hold for the majority of associations between one number and another. However, the association between any real number and its square root always yields both a positive and a negative result. For example, the two square roots of 9 are ± 3, and so we say that 9 is associated with both –3 and 3. Thus, if we write this association as $y = x^{1/2}$, then we cannot define the function $y = f(x) = x^{1/2}$. However, if we explicitly limit the values of y to the positive (or negative) roots only, then we can redefine the association as a single-valued function. Alternatively, if we square both sides to yield $y^2 = x$, we can take the association between x now as a *dependent* variable and y as an *independent* variable and define the function $x = g(x) = y^2$, for which there is only one value of x for any value of y.

The nth root of a number x may be written using the **radical** notation $\sqrt[n]{x}$, where n is the index and the $\sqrt{\ }$ sign is known as the radical. The square root of x is thus given by $\sqrt[2]{x}$, but more commonly this index is omitted and we simply write $\sqrt{x}$. By convention, use of the radical sign implies the principal or positive root. If we wish to specify explicitly the negative root, then we must write $-\sqrt{x}$. However, we may alternatively write $x^{1/2}$, which represents both positive and negative roots.

An everyday example might be the association between price and item in a supermarket. As there are in all likelihood many items that have the same price, we cannot describe this association as a function; however, the association of item with price *does* define a function, as each article has only one price. In this latter case, the domain of the function is simply a list of all items for sale in the supermarket.

Problem 2.1

State whether each of the following associations defines a function; if so, give its domain. (a) The set of car registration numbers and registered keepers. (b) The set of registered keepers and car registration numbers. (c) Periodic Table group number and element name. (d) Element name and Periodic Table group number.

2.1.1 Functions in a Chemical Context

As we have seen in the previous section, functions involve associations between numbers. However, when we work with functions in a chemical context, we have to recognize that any association involving chemical properties necessarily involves units. Consider, for example, the relation between atomic number, Z, and atomic first ionization energy, IE. While there are clearly no units associated with atomic number, the ionization energy has units of kJ mol^{-1} (although we could just as easily have chosen any other unit of energy such as eV, cm^{-1}, J, kcal mol^{-1}, and so on). In this example, there is clearly a relation between the subset of the positive integers $(1, 2, 3, 4, \ldots, 109)$, corresponding to Z, and the subset of the 109 decimal numbers corresponding to the values of IE/kJ mol^{-1}. Note that the association remains between two numbers devoid of units because, in the case of the ionization energy, we have divided IE by the units of energy chosen. For example, in the case of atomic nitrogen, where $Z = 7$, and IE/kJ mol$^{-1} = 1402.3$, there is a relation between the positive integer 7 and the decimal number 1402.3. This relation has physical meaning only for positive integers greater than or equal to 1 and less than or equal to 109 (alternatively written as $1 \leq Z \leq 109$), since each value of IE/kJ mol^{-1} is associated with only one positive integer in the subset of integers already identified. Hence, in mathematical terms, we say that the relation, or association, just described defines a function, as we have specified both the domain for which the association is valid and also checked that each number, Z, of the domain has an association with only one decimal number.

In most chemical problems we usually deal with functions that are defined in terms of a formula, in which the permitted values for the variables appearing in the formula (given by symbols) are determined by physical considerations. For example, in the case of temperature on the absolute (Kelvin) scale, negative values have no physical basis in reality.

In the next section, we explore in more detail the role that units play in the relation of formula and function.

Understanding the Role of Units: the Mathematically Correct Approach

Consider the ideal gas law, expressed in terms of the simple formula:

$$P = nRT/V \tag{2.2}$$

where the symbols have the following roles: P is the pressure, n is the amount of gas, V is the volume, T is the temperature and R is the gas constant. Each of the properties listed has associated units, and the units on both sides of equation (2.2) must be equivalent, or equal. In the SI convention, the following choices of units are common, with the equivalent combinations of base SI units also given, where appropriate: P in Pa (pascal), atm or bar, which are names defining appropriate multiples of the base units $\text{kg m}^{-1} \text{ s}^{-2}$; n in mol; V in m^3; T in K; and $R = 8.314 \text{ J K}^{-1} \text{ mol}^{-1}$, equivalent to $\text{kg m}^2 \text{ s}^{-2} \text{ K}^{-1} \text{ mol}^{-1}$. Thus, on the left side of the equation the units are $\text{kg m}^{-1} \text{ s}^{-2}$, and on the right side we have $\text{mol kg m}^2 \text{ s}^{-2} \text{ K}^{-1} \text{ mol}^{-1} \text{ K}/\text{m}^3 = \text{kg m}^{-1} \text{ s}^{-2}$, as required.

The ideal gas equation is the outcome of a model devised for understanding the properties of a gas, in which there is no interaction between the atoms or molecules occupying the volume, V. In mathematical terms, however, this ideal gas equation remains a formula until we know how to use it as a function, a key aspect of which is developed next.

Creating a Function from a Formula

As already noted, the ideal gas formula involves symbols that are associated with a value *and* its associated units. As we know that the units on the left and right sides of the formula are the same and therefore cancel, we can express each symbol as a value multiplied by appropriate units, leaving us with a relation involving new symbols that stand for numerical values in the ordinary sense of algebra. In particular, if we make the following substitutions:

$$P = p \text{ Pa}; \quad n = \bar{n} \text{ mol}; \quad R = r \text{ JK}^{-1} \text{mol}^{-1}; \quad V = v \text{ m}^3; \quad T = t \text{ K}$$

then the formula:

$$P = \frac{nRT}{V} \tag{2.3}$$

takes the form:

$$p\,\text{Pa} = \frac{\bar{n}\,\text{mol}\,r\,\text{JK}^{-1}\,\text{mol}^{-1}\,t\,\text{K}}{v\,\text{m}^3} \qquad (2.4)$$

which, on cancelling the units, becomes:

$$p = \frac{\bar{n}rt}{v} \qquad (2.5)$$

where p, $\bar{n}$, r, t and v are positive numbers, with t also permitted to take the value zero. In this case, we see that p is a function of the three variables $\bar{n}$, t and v (r is a constant). However, if the amount of gas and either temperature or volume is held constant, then there is only one independent variable and, in these circumstances, we say that p is proportional to t or that p is inversely proportional to v, respectively:

$$p \propto t \quad \text{or} \quad p \propto 1/v$$

The **constants of proportionality** are given by $\bar{n}r/v$ or by $\bar{n}rt$, respectively. In each case, we can collect our constants together and re-label them as a single constant, expressed using a new symbol. Thus, we can express the ideal gas law as:

$$p = bt \quad \text{or} \quad p = c/v$$

where $b = \bar{n}r/v$ and $c = \bar{n}rt$.

Understanding the Role of Units: the Pragmatic Approach

In the discussion above, we have seen how a formula involving chemical properties may be converted into a function, essentially by removing the units. This procedure works because the units must balance on each side of the equality ($=$) defining the formula or association. It is very easy to become confused by the distinction between formula and function and the role that units play in defining this distinction. The approach detailed above describes how to treat units in a mathematically correct fashion, but in practice the more pragmatic approach is simply to ignore the units and treat a formula describing some physical relationship as a function (for which the domain is the physically meaningful range of values for the independent variable). Consequently, we find that in most chemistry texts there is an understandable degree of mathematical looseness, which skates over this distinction between formula and function in a chemical context, and frequently results in the units being ignored. For example, it is often stated that the ideal gas law indicates that P is a function of T and V, in which $P \propto T$ and $P \propto 1/V$. The latter two statements are, of course, true, so long as it is understood that the proportionality constants carry the units of pressure divided by temperature and pressure multiplied by volume, respectively. It is not our intention here to add unnecessary levels of complexity, but it is nevertheless important to be aware of the role that units play and of the distinction between formula and function in the chemical context. We shall return to this problem of units in the next section, where we consider methods used for representing functions.

Worked Problem 2.3

The conventional wisdom in theories of molecular structure presented in organic chemistry is that the strength of a bond between identical atoms increases with increasing bond order and decreasing bond length. Thus, for example, the bond energy, BE/J mol^{-1}, of a C≡C triple bond is greater than a C=C double bond, which in turn is greater than a C−C single bond. As the bond length is inversely proportional to bond order, we can make a rough approximation that the bond energy, BE/J mol^{-1}, is inversely proportional to bond length. Thus:

$$BE/J\ mol^{-1} \propto \frac{1}{L/m}$$

Q Give the units of the proportionality constant in terms of the base SI units.

A Writing the formula in terms of base units, we have:

$$\frac{BE}{kg\ m^2\ s^{-2}\ mol^{-1}} \propto \frac{m}{L}$$

Dividing through by m on both sides gives:

$$\frac{BE}{kg\ m^3\ s^{-2}\ mol^{-1}} \propto \frac{1}{L}$$

and so we see that in order for the units to balance on both sides, the constant of proportionality must have units equivalent to the those appearing in the denominator on the left hand side, *i.e.* kg m^3 s^{-2} mol^{-1}.

A less rigorous, and somewhat more transparent, approach involves writing down the relationship between binding energy and bond length:

$$BE \propto \frac{1}{L}$$

If we replace each of the symbols BE and L by their appropriate units we have:

$$kg\ m^2\ s^{-2}\ mol^{-1} \propto \frac{1}{m}$$

and so in order to replace the proportionality symbol ($\propto$) by the equality symbol ($=$) we introduce the constant of proportionality, which must have units of kg m^2 s^{-2} mol^{-1} × m = kg m^3 s^{-2} mol^{-1}.

The movement of ions with charge ze (where z is a small positive or negative integer, and e is the fundamental unit of charge) through a solution, subject to an external electric field, E, is determined by the balance between the force arising from the electric field and the viscosity, η, of the solution. The parameter, s, termed the drift speed, gives a measure of the conductivity and is evaluated using the formula:

$$s = \frac{ezE}{6\pi\eta a}$$

where a is the effective radius of the ion. Given that the units of e, E, η and a are C, V m^{-1}, kg m^{-1} s^{-1} and m, respectively, give the units of s (remember that C V $=$ J, and J $=$ kg m^2 s^{-2}).

2.2   Representation of Functions

Functions of a single variable, involving a relation between two sets of numbers, may be expressed in terms of a table (expressing an association), formula, prescription or graphical plot. For functions of two independent variables (see below), the preferred representations are formula, prescription or graphical plot; for three or more variables, a formula or prescription is the only realistic representation.

2.2.1 Tabular Representations of Functions of a Single Variable

The function $y = g(x)$, where $g(x) = 2x + 1$, with the domain consisting of the *integers* from −5 to 5, can most easily be expressed in tabular form (see Table 2.1). For each value of x there exists one value of y.

Table 2.1   The function $g(x) = 2x + 1$, with the domain consisting of the integers from −5 to 5 expressed in tabular form

x	−5	−4	−3	−2	−1	0	1	2	3	4	5
$g(x)$	−9	−7	−5	−3	−1	1	3	5	7	9	11

It is clear that there are 11 numbers (elements) in the domain. However, it is not possible to present the function $f(x) = 2x + 1$ with the domain of all real numbers from −5 to 5 in tabular form, as there is an infinite number of elements in the domain. The formula $f(x) = 2x + 1$ is the most effective non-graphical way of specifying this function, with the domain as specified above.

2.2.2 Graphical Representations of Functions of a Single Variable

For the function $y = f(x)$, each ordered pair of numbers, (x,y), can be used to define the coordinates of a point in a plane, and thus can be represented by a graphical plot, in which the **origin**, O, with coordinates $(0,0)$, lies at the intersection of two perpendicular axes. A number on the horizontal x-axis is known as the **abscissa**, and defines the x-coordinate of a point in the plane; likewise, a number on the (vertical) y-axis is known as the **ordinate**, and defines the y-coordinate of the point. Thus, an arbitrary point (x,y) in the plane lies at a perpendicular distance $|x|$ from the y-axis and $|y|$ from the x-axis. If $x > 0$, the point lies to the right of the y-axis; if $x < 0$, it lies to the left. Similarly, if $y > 0$, the point lies above the x-axis, and if $y < 0$, it lies below (see Figure 2.1).

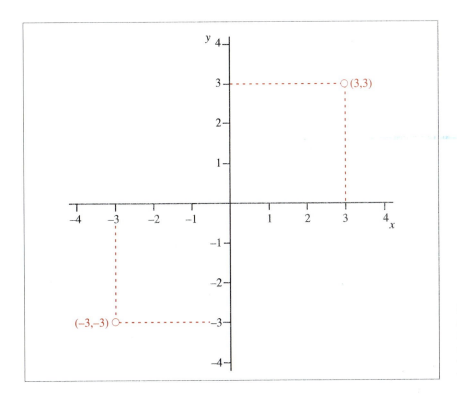

Figure 2.1 The Cartesian coordinate system used to represent the points (3,3) and (−3,−3) in the plane defined in terms of coordinates referenced to the origin (0,0)

For functions such as $y = g(x)$, where $g(x) = 2x + 1$, the most appropriate type of plot is a **point plot** if the domain is limited to integers lying within some range. Figure 2.2 displays such a plot for this function, which is defined only at the 11 values of x in its domain (indicated by small open circles).

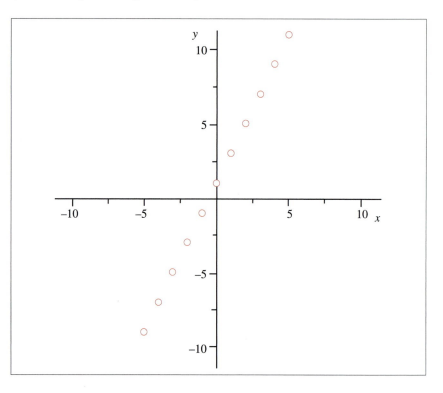

Figure 2.2 A point plot illustrating the values of the function $y = g(x)$ in its domain [−5,5]

Strictly speaking, it is not appropriate to connect the points with a line, as this would imply that the function is defined at points other than at the integers from −5 to +5. However, in some instances, as we shall see below, it may be appropriate to connect the data points with straight line segments in order to guide the eye, but this has no mathematical significance. In contrast, the **line plot** of the function $y = f(x) = 2x + 1$ is created by taking a sufficient number of points in its domain, to enable a smooth curve to be drawn (Figure 2.3). In this case, it would be similarly misleading to represent this plot as a series of discrete points, no matter how small the gap between adjacent points. The only way of correctly representing this function is as a smoothly varying line plot, but, of course, in practice we recognize that the logistics of generating a graphical line plot involve arbitrarily selecting discrete points within the domain and then joining the points. This applies equally whether we are drawing the plot by hand or using a computer plotting program.

A Chemical Example of a Point Plot

Consider the association between the atomic first ionization energy, IE, and the atomic number Z (a positive integer). It is convenient to use the

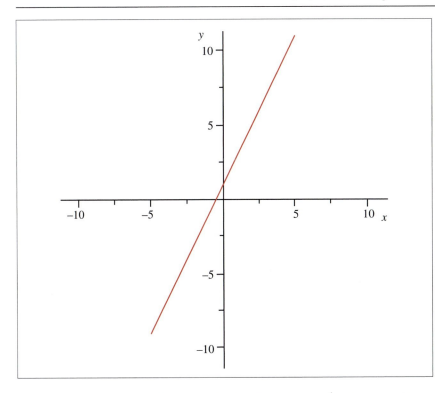

Figure 2.3 Line plot of the function $y = f(x) = 2x + 1$

electronvolt unit, eV, where $1 \text{ eV} = 96.485 \text{ kJ mol}^{-1}$. As there is no formula to express this association, we present the function first in the form of a table, and then as a point plot; in both representations we take as domain the Z values for the first 18 elements. Since units have to be removed in order to define a function, we consider in Table 2.2 the association of Z with IE/eV.

Table 2.2   Atomic number and ionization for the first 18 elements

Z	1	2	3	4	5	6	7	8	9
IE/eV	13.6	24.6	5.4	9.3	8.3	11.3	14.5	13.6	17.4
Z	10	11	12	13	14	15	16	17	18
IE/eV	21.6	5.1	12.8	6.0	8.2	10.5	10.4	13.0	15.8

The data in the table are now displayed in graphical form as a point plot (or scatter plot) in Figure 2.4, with points defined by the number pairs (Z, IE/eV). In this procedure, Z is specified as abscissa (x-axis) and IE/eV as ordinate (y-axis).

For every value of Z there is clearly only one value for the ionization energy, which establishes a function with domain given by the set of the first 18 positive integers, a subset of the atomic numbers of the 109 elements in the Periodic Table. In this example, the fact that the data

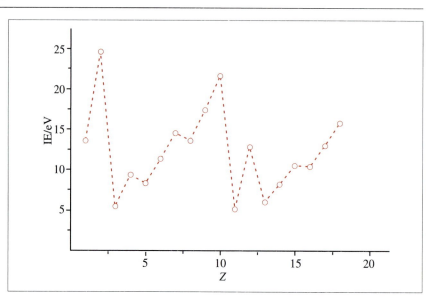

Figure 2.4 A point plot displaying atomic number *versus* ionization energy in eV

points are connected by dashed straight-line segments has no mathematical significance: it simply acts as a visual aid to improve the display of the trends in IE/eV values.

There are many situations where we are unable to provide a formula that relates one chemical property with another, even though, intuitively, one may be expected. Thus, in the example given above, it is not possible to construct a simple model, based on a formula, that relates Z to IE/eV for atoms containing more than one electron (although a simple relationship does exist for one-electron species, if we ignore relativistic effects). However, regardless of whether a particular problem is as intractable as this one, we can only enhance our understanding of chemistry by using the mathematical tools at our disposal to develop new models to crack particular chemical "nuts"! As an example of this kind of model development, we now consider some pressure/volume data for a real gas in order to test the ideal gas law, derived from the Boyle model, and to see how we can refine the law to find a better "fit" to our data.

In Section 2.1.1 we saw that, for an ideal gas, the numerical values of the pressure, p, and volume, v, are related according to $p \propto 1/v$, or $p = c/v$, where $c = \bar{n}rt$ (a constant). We can now explore how well the ideal gas law works for a real gas by considering experimental data[1] for 1 mol of CO_2 at $T = 313$ K. The ideal gas law suggests that pressure is inversely proportional to the volume and so in the first two rows of Table 2.3 we present the variation of p with $1/v$ for the experimental data (note that the working units for the pressure and volume in this case are atm and dm^3, respectively). In the third row, we show values for $1/v_B$, obtained using the ideal gas equation, where, in this case, the constant of proportionality $c = 25.6838$ at $T = 313$ K. The data in the fourth and fifth rows derive from a refinement to the model, discussed below.

Table 2.3 A comparison of experimental p versus $1/v$ data for 1 mol of CO_2 at 313 K with values for $1/v$ generated from the ideal gas law, from a fit to the van der Waals equation, and from the van der Waals equation but using the book values[2] for the constants a and b (see text for details)

$p = P$/atm	1	10	50	100	200	500
$1/v = 1/(V/\text{dm}^3)$	0.0392	0.4083	2.6316	14.300	19.048	22.727
$1/v_B$	0.0389	0.3893	1.9467	3.8934	7.7867	19.467
$1/v_{\text{vdW,fit}}$	0.0391	0.4010	2.3294	7.402	19.175	22.730
$1/v_{\text{vdW,book}}$	0.0391	0.4048	2.5189	11.249	14.184	16.835

It should be quite obvious that, although the model provided in the form of the ideal gas law does a reasonable job at lower pressures, it rapidly deviates as the pressure increases and the volume decreases. We can see this more clearly in Figure 2.5, where we compare the real data with that derived from the ideal gas law in a scatter plot of p *versus* $1/v$. We can see from our plot that the experimental data, shown as solid circles, are modelled reasonably well by a linear (straight line) function, but only for pressures less than 50 atm. The Boyle model is clearly of limited applicability in this case.

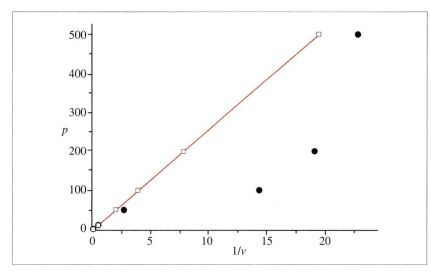

Figure 2.5 Plot of p *versus* $1/v$, assuming the Boyle model (*open box symbols*). Experimental data for CO_2 at 313 K are shown as *solid circles*

Improving on the Boyle Model

An example of a model equation for a real gas is provided by the van der Waals equation:

$$P = \frac{RT}{V - b} - \frac{a}{V^2} \qquad (2.6)$$

in which some essence of non-ideality is included through the two parameters a and b. Values of a and b for CO_2 can be obtained by fitting the experimental data to this model expression. For the experimental data set given in Table 2.3, we obtain values for a and b of 2.645 atm dm^6 mol^{-2} and 3.025×10^{-2} dm^3 mol^{-1}, respectively. The book values for a and b are 3.592 atm dm^6 mol^{-2} and 4.267×10^{-2} dm^3 mol^{-1}, respectively.[2] The differences arise from the limited number of data points available to us, but we can see that, in spite of this, our fitted values for a and b are of the same order of magnitude as the book values. If we now compare a scatter plot of p versus $1/v_{\mathrm{vdW,fit}}$, using our fitted values for a and b (Figure 2.6), we see that, although the fit to our experimental data is really quite good in both the low- and high-pressure regions, it is quite poor in the region of the critical point where we have a **point of inflection** on our plot (see Chapter 4). Using the book values for the van der Waals constants gives a reasonable fit below 50 atm, but increasingly poorer fits to higher pressures; however, the fit in the critical region is much better than that achieved from our fitted values for a and b. While we have not been able to construct a model that fits the experimental data perfectly, it is a considerable improvement on the Boyle model and allows some insight into what factors might be causing the deviation from ideal gas behaviour. Furthermore, our model provides a starting point for further refinements, which might focus, for example, on improving the model for different regions of the domain (such as in the critical region) or even taking a different approach altogether, such as looking for a polynomial function in $1/v$ that has a more extended validity (see Chapter 1 in Volume 2).

The point at which the phase boundary between liquid and gas phases disappears is known as the critical point.

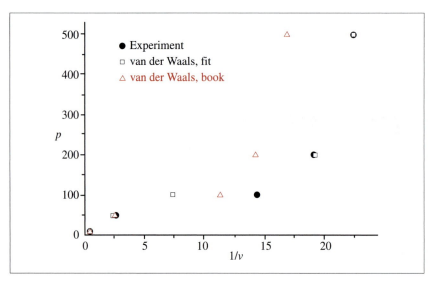

Figure 2.6 van der Waals (*open symbols*) and experimental data (*solid circles*), used in a pressure *versus* volume plot for CO_2

2.2.3 Representing a Function in Terms of a Prescription

The simplest function defined in terms of a **prescription** (such functions are sometimes termed piecewise functions) is the **modulus function**, $f(x) = |x|$ defined as follows:

$$f(x) = \begin{cases} x, x \geq 0 \\ -x, x < 0 \end{cases} \tag{2.7}$$

a plot of which is given in Figure 2.7.

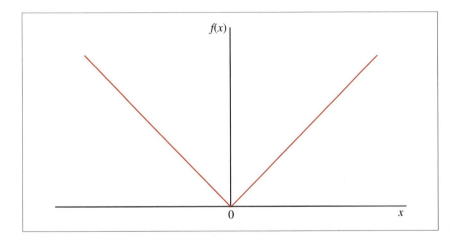

$f(x)$

0 x

Figure 2.7 The modulus function $y = |x|$

The modulus function in equation (2.7) is an example of a function displaying a "kink" at the origin. In this case it is necessary to split the domain into two subintervals, in each of which the formula takes a different form. Further examples of this type of behaviour are described in Chapter 3.

Problem 2.3

(a) Give the prescription for the function $f(x) = |x - 1|$, and sketch its form.
(b) Sketch the unit pulse function with the prescription

$$g(x) = \begin{cases} 0, & x < 1 \\ 1, & 1 \leq x < 2 \\ 0, & x \geq 2 \end{cases}$$

Note: the unit pulse function is used for modelling NMR spectra.

Prescription Functions in Chemistry

Functions, specified in the form of a prescription, are required when describing properties of chemical systems that undergo phase changes. For example:

• The function describing the change in entropy, as a function of temperature, involves the use of a prescription that contains a formula specific to a particular phase. At each phase transition temperature the function suffers a finite jump in value because of the sudden change in thermodynamic properties. For example, at the boiling point T_b the sudden change in entropy is due to the latent heat of evaporation (see Figure 2.8).

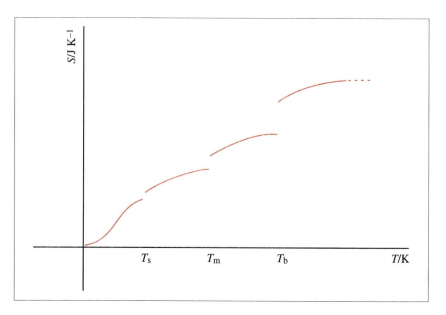

Figure 2.8 A plot of the function describing the change in absolute entropy as a function of temperature. The discontinuities occur at phase changes

• The function describing the change in equilibrium concentration of a given species following a sudden rise in temperature (in a so-called temperature jump experiment), has two parts, corresponding to times before and after the temperature jump (Figure 2.9).

Transcendental functions are mathematical functions which cannot be specified in terms of a simple algebraic expression involving a finite number of elementary operations $(+, -, \div, \times)$. By definition, functions which are not transcendental are called algebraic functions.

2.3   **Some Special Mathematical Functions**

There are many different kinds of function in mathematics, but in this chapter we shall restrict the discussion to those **transcendental functions**, such as exponential, logarithm and trigonometric functions, that have widespread use in chemistry.

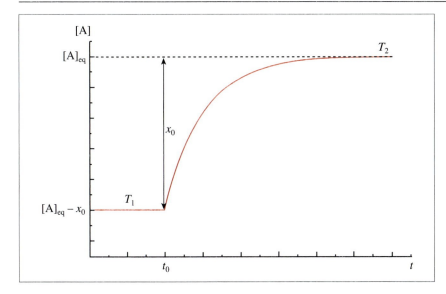

Figure 2.9 The exponential relaxation of the equilibrium concentration to a new equilibrium concentration following a sudden temperature jump from T_1 to T_2

2.3.1 Exponential Functions

In Chapter 1 we saw that there are 2^n spin states for n equivalent protons, where the physics of such systems requires that $n \geq 1$. If we now change the name of the independent variable from n to x, we can define the function $y = f(x) = 2^x$ with a domain, for example, initially restricted to the integer values -4 to $+4$. We have displayed this function as a scatter plot in Figure 2.10. If we now extend the domain to any real value for x, we can define the **exponential** function $y = g(x) = 2^x$, with **base** equal to 2, part of which is displayed in Figure 2.10 as the full line plot.

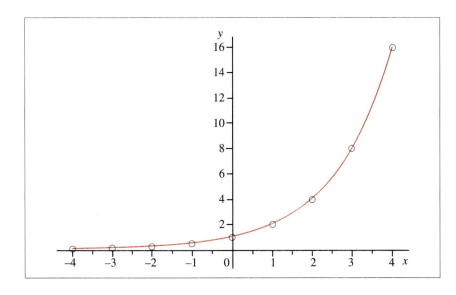

Figure 2.10 Scatter and line plots of the functions $y = f(x) = 2^x$, domain $x = -4, -3, -2, -1, 0, 1, 2, 3, 4$ (*open circles*) and $y = g(x) = 2^x$, domain **R** (*full line*)

In chemistry, in addition to base 2, which we meet rather infrequently, we also encounter exponential functions with base 10 in, for example, relating pH to the activity, a, of hydronium ions in aqueous solutions, using the formula $a = 10^{-\text{pH}}$. However, the base most commonly encountered is provided by the unexpectedly strange, irrational number e, which has the value 2.718 281 828 ... This base arises when describing growth and decay processes in chemistry, *e.g.* in kinetics, where changes in concentration with respect to time are the focus of attention, and in quantum chemistry, where we are interested in the changes in the probability density function for finding an electron at a particular point in space. In a mathematical context, however, e defines the base of the natural logarithm function (see below), and also has a major role in calculus (Chapter 4).

In comparing exponential functions with different bases, the larger the base, the more rapidly the value of the function increases with increasingly large positive values of x, and decreases with increasingly negative values of x. The value of y at $x = 0$ is unity, irrespective of the choice of base. Regardless of the choice of base, exponential functions display a **horizontal asymptote** at $y = 0$: as x takes on increasingly large negative values, the curve approaches the line $y = 0$ but never crosses it. We explore the limiting behaviour of functions in more detail in Chapter 3.

Two Chemical Examples

- In modelling the vibrational "umbrella" mode for ammonia, the potential energy function $V = \frac{1}{2}kx^2 + be^{-cx^2}$ is commonly used, where b and c are constants (see Figure 2.11).

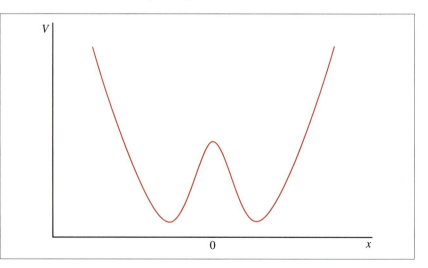

Figure 2.11 A plot of the potential energy function $V = \frac{1}{2}kx^2 + be^{-cx^2}$, using appropriate values of b and c, to describe the umbrella motion in ammonia

- The number of molecular species, n_i, occupying a given energy state, ε_i, is estimated using the Boltzmann distribution function:

$$n_i = n_0 e^{-(\varepsilon_i - \varepsilon_0)/kT} \tag{2.8}$$

where n_0 is the number of species in the lowest energy state, k is the Boltzmann constant, T the temperature and the suffix i takes values 0, 1, 2, 3, . . . Since this function has the domain of positive integers, it can only be visualized graphically using a point plot (see Figure 2.12).

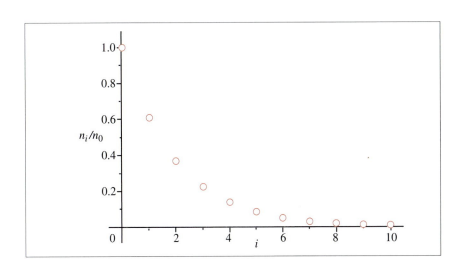

Figure 2.12 A scatter plot of the Boltzmann distribution showing the fractional population of energy levels at a given temperature, T

Problem 2.4

If the vibrations of carbon dioxide are assumed to be harmonic, then the energy states are equi-spaced with $\varepsilon_i = (i + \frac{1}{2})hv$, where v is one of the four vibration frequencies. (a) Use the Boltzmann distribution function to show that $\frac{n_i}{n_0} = e^{-ihv/kT}$. (b) Tabulate the values of $\frac{n_i}{n_0}$ to four significant figures for the Raman-active vibration with $v = 0.4032 \times 10^{14}$ s^{-1}, taking $i = 1, 2, 3, 4, 5$, $T = 300$ K, $k = 1.381 \times 10^{-23}$ J K^{-1} and $h = 6.626 \times 10^{-34}$ J s.

2.3.2 Logarithm Functions

Logarithm functions appear widely in a chemical context, for example in studying:

* The thermodynamic properties of an ensemble of atoms or molecules.
* The model equations for first- and second-order kinetics.
* The temperature dependence of equilibrium constants.

Defining the Logarithm Function

If $y = a^x$ (a is the base), then we define the logarithm to the base a of y to be x, *i.e.*:

$$\log_a y = x \tag{2.9}$$

It follows that:

$$a^{\log_a y} = a^x = y \tag{2.10}$$

Properties of Logarithms

Given two numbers y_1, y_2, such that $y_1 = a^{x_1}$ and $y_2 = a^{x_2}$, we have from the definition:

$$\log_a(y_1 y_2) = \log_a(a^{x_1 + x_2}) = x_1 + x_2 \tag{2.11}$$

However, again from the definition, we have $\log_a y_1 = x_1$ and $\log_a y_2 = x_2$, and hence:

$$\log_a(y_1 y_2) = \log_a y_1 + \log_a y_2 \tag{2.12}$$

By a simple extension of this argument, we find that:

$$\log_a(y^n) = n \log_a y \tag{2.13}$$

Note that this applies equally if the index is negative; thus:

$$\log_a(y^{-n}) = -n \log_a y \tag{2.14}$$

Similarly, by using the laws of indices and the defining relations for logarithms above, we have:

$$\log_a\left(\frac{y_1}{y_2}\right) = \log_a\left(\frac{a^{x_1}}{a^{x_2}}\right) = \log_a(a^{x_1 - x_2}) = x_1 - x_2$$
$$= \log_a y_1 - \log_a y_2 \tag{2.15}$$

Finally, to convert the logarithm from base a to base b, we can use the initial equality:

$$y = a^{x_1} = b^{x_2} \tag{2.16}$$

to give:

$$\log_b y = \log_b(a^{x_1}) = x_1 \log_b a \text{ and } x_1 = \log_a y \tag{2.17}$$

and hence:

$$\log_b y = \log_a y \log_b a \tag{2.18}$$

A Convention

Logarithm functions with the bases e and 10 are usually designated by ln and log, respectively.

Worked Problem 2.4

Q Using appropriate properties of logarithms listed above, and without the aid of a calculator, evaluate log 1−log 100.

A $\log 1 - \log 100 = \log \frac{1}{100} = \log 100^{-1} = -1 \times \log 100 = -2$.

Problem 2.5

(a) Express the following in terms of log 2: (i) log 4; (ii) log 8; (iii) log 6−log 3; (iv) ln 8; (v) ln $\frac{1}{2}$. *Hint*: for part (iv), you will need to convert from base e logs to base 10 logs using equation (2.18).
(b) Simplify the following expressions: (i) log 2 + log 3; (ii) ln 3−ln 6.

Problem 2.6

(a) Given that pH$= -\log a_H$, where a_H is the activity of hydronium ions, derive an expression for pH in terms of ln a_H.
(b) In electrochemistry the standard electromotive force, E°, of a cell is related to the equilibrium constant, K, for the cell reaction according to the formula:

$$E^{\circ} = -\frac{RT}{nF}\ln K$$

where n is the number of electrons involved and F is the Faraday constant. Find an equivalent expression in terms of log K.
(c) The strength of a weak monobasic acid HA, with dissociative equilibrium constant K, is measured in terms of a value of pK, where p$K = -\log K$. Find the value of K, to four sig. figs., for ethanoic acid, given that p$K = 4.756$.

2.3.3 Trigonometric Functions

Consider the right-angled triangle shown in Figure 2.13. The basic **trigonometric functions sine** and **cosine**, given the names sin and cos, respectively, are defined using the ratios of the side-lengths of a right-angled triangle as:

$$\sin\theta = \frac{BC}{AB} \qquad (2.19)$$

and

$$\cos\theta = \frac{AC}{AB} \qquad (2.20)$$

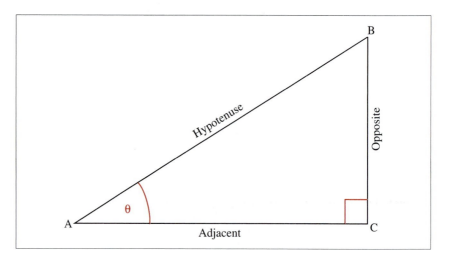

Figure 2.13 A right-angled tri-angle, with angle BAC specified as θ

The sides of a right-angled triangle are referred to as the adjacent or base (AC); opposite or perpendicular (BC); and hypotenuse (AB), opposite the right angle. The **tangent** of the angle θ is given by the quotient of $\sin\theta$ and $\cos\theta$:

$$\tan\theta = \frac{\sin\theta}{\cos\theta} = \frac{BC}{AC} \qquad (2.21)$$

The Question of Angle

Figure 2.14 shows a circle of radius r and an arc (a portion of the circumference) of length s, subtended by the angle θ. There are two basic measures of the angle θ: the first, and probably more familiar, is the **degree**. The angle θ has the value of one degree if the arc-length s is equal to one 360th of the circumference of the circle; and so a complete revolution corresponds to 360 degrees, with half a revolution corresponding to 180 degrees and a quarter to 90 degrees (a right angle). The second measure of angle is the **radian**; one radian is the angle made when the arc length is equal to the radius of the circle; in other words, it is defined in terms of the ratio of arc length to radius, *i.e.* $\theta = s/r$. As the circumference of the circle is equal to $2\pi r$, it follows that there must be 2π radians in one complete revolution, π radians in half a revolution and $\pi/2$ in a quarter revolution. Since π radians is equivalent to 180 degrees, we can see that one radian must equal $180/\pi = 57.296$ degrees (to 5 sig. figs).

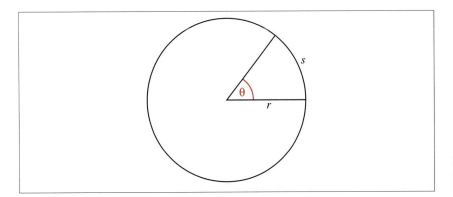

Figure 2.14   A circle of radius r and an arc of length s, subtended by the angle θ

Angle Measure in SI Units

Since radians are defined in terms of the ratio of two lengths, the values associated with an angle carry no units and are said to be dimensionless. Similarly, the degree measure of angle is also dimensionless. We can reinforce this by remembering that the sine or cosine of an angle, whether measured in degrees or radians, is defined as the ratio of the lengths of two sides of a triangle. The dimensions of sine or cosine must then cancel, which, in turn, implies that the angle itself is dimensionless. However, in order to indicate which form of angle measure is in use, it is common practice to attach the SI symbol "rad" (as a quasi unit), or to place a small circle as a superscript to indicate degrees. For example, we have $\theta = \pi/2$ rad or $\theta = 90°$ or, equivalently, $\theta/\text{rad} = \pi/2$ or $\theta/° = 90$.

Sign Conventions for Angles and Trigonometric Functions

The geometric definition given above for the trigonometric functions, in terms of the ratio of the sides of a right-angled triangle, restricts the angles to values in the range $0°$ to $90°$ or, alternatively, to:

$$0 \leqslant \theta \leqslant \frac{\pi}{2} \text{ rad}$$

The definition of angle may be broadened, however, by considering the location of a point (x,y) on the circumference of a circle, with centre at the origin of a **Cartesian** xy-coordinate system (sometimes referred to as a

rectangular coordinate system) (see Figure 2.15). The line joining the point on the circle to the origin is of length r, and equal to the radius of the circle.

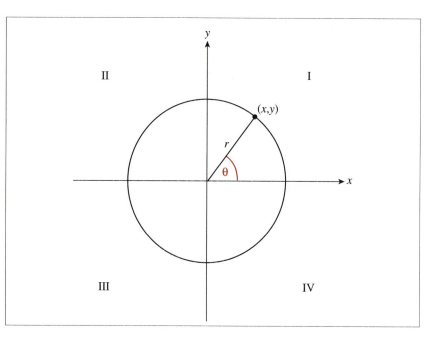

Figure 2.15 The angle θ represented in terms of a circle placed on a Cartesian coordinate system

A zero value for the angle corresponds to the point lying on the positive x-axis. The angle increases in a positive sense as the point circulates in an anticlockwise direction; circulation in a clockwise sense is indicated by a negative value of θ. Thus, for example, $\theta = -\frac{\pi}{4}$ is equivalent to $\theta = \frac{7\pi}{4}$ (see Figure 2.16).

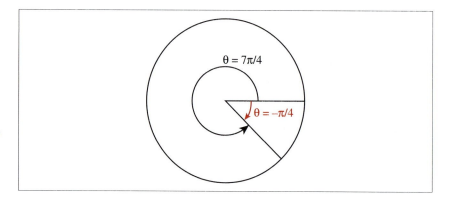

Figure 2.16 Positive values for an angle are generated by rotation in an anti-clockwise sense, while negative values imply clockwise rotation

We can now redefine the trigonometric functions in terms of the radius r, and the coordinates x and y, of a point on the circle:

$$\sin\theta = \frac{y}{r}, \quad \cos\theta = \frac{x}{r} \text{ and } \tan\theta = \frac{y}{x} \quad (2.22)$$

These definitions are not in conflict with those given earlier, but now allow for all angles; for example, angles lying in the range $90° < \theta < 180°$ correspond to negative values for x and positive values for y, whilst those in the range $270° < \theta < 360°$ correspond to positive values for x and negative values for y. We can also see how the signs of the values of the trigonometric functions depend upon which of the four quadrants of the circle the point lies in (see Figure 2.15 and Table 2.4). For example, in quadrant II, where $90° < \theta < 180°$, and where x is negative and y positive (and remembering that r is always positive): $\sin\theta > 0$, $\cos\theta < 0$ and $\tan\theta < 0$.

Table 2.4 The signs of the trigonometric functions sin, cos and tan in each of the four quadrants shown in Figure 2.14

Quadrant	Function		
	sin	cos	tan
I	+	+	+
II	+	−	−
III	−	−	+
IV	−	+	−

Special Values for Trigonometric Functions

There are only a few special cases where trigonometric functions have exact values, all of which are obtained easily without reference to tables or resorting to the use of a calculator. For example, $\sin(\pi/4)$ is calculated from the definition of the sine function and use of Pythagoras. An angle of $\theta = \pi/4$ (or $\theta = 45°$) requires the magnitudes of x and y to be equal, which implies the length of the hypotenuse (given by r) to be a factor of $\sqrt{2}$ larger than either x or y. Thus, if $\sin\theta = y/r$, then $\sin(\pi/4) = \frac{1}{\sqrt{2}}$. Table 2.5 lists some of the special values for the sine, cosine and tangent functions.

Pythagoras' Theorem states that, for a right-angled triangle, the square of the length of the hypotenuse is equal to the sum of the squares of the lengths of the remaining two sides. Using Figure 2.13 for reference, we can write this more succinctly as $AB^2 = AC^2 + BC^2$.

Table 2.5 Some special values for the sin, cos and tan functions

	$\theta = 0$	$\pi/6$	$\pi/4$	$\pi/3$	$\pi/2$	π	$3\pi/2$	2π
$\sin\theta$	0	$\dfrac{1}{2}$	$\dfrac{1}{\sqrt{2}}$	$\dfrac{\sqrt{3}}{2}$	1	0	−1	0
$\cos\theta$	1	$\dfrac{\sqrt{3}}{2}$	$\dfrac{1}{\sqrt{2}}$	$\dfrac{1}{2}$	0	−1	0	1
$\tan\theta$	0	$\dfrac{1}{\sqrt{3}}$	1	$\sqrt{3}$	∞	0	$-\infty$	0

An isosceles triangle is a triangle with at least two sides of equal length and with two equal angles. The name derives from the Greek *iso* (same) and *skelos* (leg). A triangle having all sides of equal length is called an **equilateral** triangle, but because it has two sides of equal length it is also a special case of an isosceles triangle. A triangle with no equal length sides is called a **scalene** triangle.

Problem 2.7

(a) Given that the PH_2 radical has a "V" shape in its ground state, with an HPH angle of 123° and a P–H bond length of 140 pm in its ground state, calculate the H–H distance. (b) Given that the bond angle and P–H bond length change to 107° and 102 pm, respectively, in the first electronic excited state, calculate the change in the H–H distance.

Hint: you may find it helpful to draw an **isosceles** triangle, and drop a perpendicular from the **P** atom to the line joining the two protons.

Reciprocal Trigonometric Functions

Three further trigonometric functions, **cosecant**, **secant** and **cotangent**, are provided by the reciprocals of the basic functions:

$$\operatorname{cosec} \theta = \frac{1}{\sin \theta}, \ \sec \theta = \frac{1}{\cos \theta}, \ \cot \theta = \frac{1}{\tan \theta} = \frac{\cos \theta}{\sin \theta} \tag{2.23}$$

Domains and Periodic Nature of Trigonometric Functions

Thus far we have considered angles ranging from 0 to 2π (0 to 360°), but we can further extend this range by allowing additional complete rotations about the origin. Each additional rotation, anticlockwise or clockwise, adds or subtracts 2π to or from the angle, with the value of the sine and cosine trigonometric functions simply repeating with each full rotation. The tangent function repeats every half rotation. Thus, for the angles $\theta \pm 2n\pi$, where $n = 0, 1, 2, 3, \ldots$:

$$\sin(\theta \pm 2n\pi) = \sin \theta \text{ and } \cos(\theta \pm 2n\pi) = \cos \theta \tag{2.24}$$

and for the angles $\theta \pm n\pi$:

$$\tan(\theta \pm n\pi) = \tan \theta \tag{2.25}$$

Plots of the three trigonometric functions are shown in Figure 2.17.

Many-to-one functions are those for which more than one value of x is associated with one value of $f(x)$. Thus, for $f(x) = x^2$, the numbers ± 2 are both associated with the number 4, and so this function is 2:1.

Functions having a property $f(x \pm a) = f(x)$ are known as **periodic functions** with a **period** a, and are said to be **many-to-one functions**. In the examples given above, the period for the sine and cosine functions is 2π, while that for the tangent function is π.

We can see from Table 2.5 and Figure 2.17 that the sine and cosine functions both have as domain the set of real numbers. The domains of the tangent and reciprocal trigonometric functions are different, however,

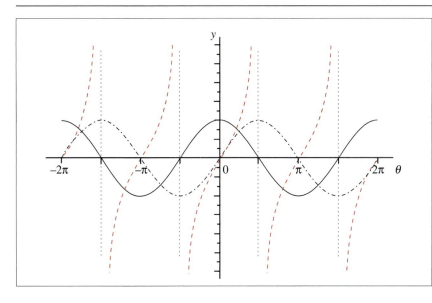

Figure 2.17 Plots of the trigonometric functions $\sin\theta$ (*dot-dash line*), $\cos\theta$ (*full line*), and $\tan\theta$ (*dashed line*) for $-2\pi \leq \theta \leq 2\pi$. The principal branch of each function is shown by the *thick lines*. The *dotted vertical lines* at odd multiples of $\pi/2$ indicate the points of discontinuity in the tangent function at these values of θ

because we must exclude values of θ for which the denominator of the defining formula is zero. Thus, for example, since $\cos\theta = 0$ for $\theta = (2n-1)\pi/2$, where n is any integer (including zero), the domains for the secant (sec) and tangent (tan) functions consist of the set of real numbers, with the exclusion of $\theta = (2n-1)\pi/2$ with n defined as above. For the tan and sec functions, the lines at $\theta = (2n-1)\pi/2$ are known as **vertical asymptotes**, because the curves of the respective functions approach these lines without ever crossing them (see Figure 2.17). In some situations it is necessary to limit the domains so that the functions are so-called 1:1 functions (as opposed to many-to-one). The **principal branches** for the sine, cosine and tangent functions, chosen by convention to define them as 1:1 functions, are:

$$\sin\theta : \quad -\frac{\pi}{2} \leq \theta \leq \frac{\pi}{2} \tag{2.26}$$

$$\cos\theta : \quad 0 \leq \theta \leq \pi \tag{2.27}$$

$$\tan\theta : \quad -\frac{\pi}{2} \leq \theta \leq \frac{\pi}{2} \tag{2.28}$$

Problem 2.8

Give the domains of the cotangent (cot) and cosecant (cosec) functions.

Important Identities Involving Trigonometric Functions

The Addition Formulae

Expressions for the sine and cosine of the sum or difference of two angles are given by the following formulae:

$$\sin (A \pm B) = \sin A \cos B \pm \sin B \cos A \tag{2.29}$$

$$\cos (A \pm B) = \cos A \cos B \mp \sin A \sin B \tag{2.30}$$

Thus, for example, in the discussion of periodicity above, and with the use of Table 2.5, we see that:

$$\sin (\theta + 2\pi) = \sin \theta \cos 2\pi + \sin 2\pi \cos \theta = \sin \theta \tag{2.31}$$

$$\cos (\theta + 2\pi) = \cos \theta \cos 2\pi - \sin\theta \sin 2\pi = \cos \theta \tag{2.32}$$

Useful Identities

$$\cos^2 A + \sin^2 A = 1 \tag{2.33}$$

$$\cos 2A = \cos^2 A - \sin^2 A \tag{2.34}$$

$$\sin 2A = 2 \cos A \sin A \tag{2.35}$$

where the expressions $\cos^2 A$ and $\sin^2 A$ mean $(\cos A)^2$ and $(\sin A)^2$, respectively. All the other identities that we may need follow from these three identities and the addition formulae. For example:

$$\cos 3A = \cos (2A + A) = \cos2A \cos A - \sin 2A \sin A$$
$$= \cos^3 A - \sin^2 A \cos A - 2 \sin^2 A \cos A \tag{2.36}$$
$$= \cos^3 A - 3 \sin^2 A \cos A$$

but since $\sin^2 A = 1 - \cos^2 A$, we can rewrite this as:

$$\cos3A = 4 \cos^3 A - 3 \cos A \tag{2.37}$$

Further Important Properties of Trigonometric Functions

Since negative angles arise when using trigonometric functions, it is important to establish how, for example, $\sin(-\theta)$ is related to $\sin\theta$. The periodicity of the sine function yields the equality:

$$\sin(-\theta) = \sin(2\pi - \theta) \tag{2.38}$$

and, so using the sine addition rule, we obtain:

$$\sin(-\theta) = \sin 2\pi \cos \theta - \sin \theta \cos 2\pi = -\sin \theta \tag{2.39}$$

Problem 2.9

Repeat the above example for cos(−θ) and tan(−θ), remembering the definition of tan θ.

2.3.4 Exponential Functions with Base e Revisited

The hyperbolic sine and cosine functions **sinh** x and **cosh** x are defined in terms of the sum and difference of the exponential functions e^x and e^{-x}, respectively:

$$\sinh x = \frac{1}{2}(e^x - e^{-x}) \tag{2.40}$$

$$\cosh x = \frac{1}{2}(e^x + e^{-x}) \tag{2.41}$$

and have the graphical forms depicted in Figure 2.18.

The other **hyperbolic functions**, **tanh**, **cosech**, **sech** and **coth**, defined in terms sinh and cosh, are the hyperbolic analogues of the functions tan, cosec, sec and cotan, and are defined as follows:

$$\tanh x = \frac{\sinh x}{\cosh x}, \quad \operatorname{cosech} x = \frac{1}{\sinh x},$$

$$\operatorname{sech} x = \frac{1}{\cosh x}, \quad \coth x = \frac{\cosh x}{\sinh x} \tag{2.42}$$

The coth and tanh functions play an important role in the modelling of the magnetic behaviour of transition metal complexes.

Problem 2.10

Use the definitions for sinh x and cosh x to show that: (a) (i) sinh x + cosh $x = e^x$; (ii) sinh x−cosh $x = -e^{-x}$ (b) (i) $\cosh^2 x - \sinh^2 x = 1$; (ii) $\cosh^2 x + \sinh^2 x = \cosh 2x$; (iii) sinh $2x = 2$sinh x cosh x.

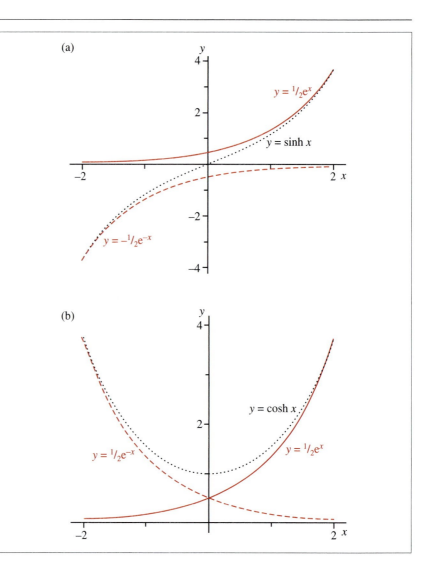

Figure 2.18 Plots of the hyperbolic functions (a) $y = \sinh x$ and (b) $y = \cosh x$ compared with the exponential functions $y = \frac{1}{2}e^x$, $y = \frac{1}{2}e^{-x}$ and $y = -\frac{1}{2}e^{-x}$

Symmetric and Antisymmetric Functions

Functions having the property $f(-x) = f(x)$ are called **symmetric**, or **even**, functions, whilst those having the property $f(-x) = -f(x)$ are called **antisymmetric** or **odd** functions. In our discussion of trigonometric and hyperbolic functions, we have encountered a number of examples of functions that fall into one or other of these categories, as well as some that fall into neither. Symmetric and antisymmetric functions are so called because they are symmetric or antisymmetric with respect to reflection in the y-axis. A close look at Figure 2.17 shows that, since $\cos\theta = \cos(-\theta)$, and $\sin\theta = -\sin(-\theta)$, the cos and sin functions are symmetric and antisymmetric, respectively. Likewise, we can classify the cosh and sinh functions as symmetric and antisymmetric, respectively (see Figure 2.18). The exponential functions displayed in Figure 2.18 are neither symmetric nor antisymmetric. In Chapter 6 we shall meet these ideas again, when we consider the integration of functions having well-defined symmetry, a feature that has important applications in quantum mechanics where we consider the physical significance of whether certain integrals involving wave functions of atoms and molecules are zero or non-zero.

The Product Function $x^2 e^{-x}$

The function $y = x^2 e^{-x}$ is a product of two functions, x^2 and e^{-x}; the former *increases* rapidly with increasing x, but the latter *decreases* even more rapidly with increasing x. The result is that the value of the product function, which is initially dominated by the quadratic term, will quite rapidly be overcome by the exponential term as x increases. In fact this is true regardless of the degree of the power term: it does not matter whether we consider the function $y = x^2 e^{-x}$ or $y = x^{20} e^{-x}$ or $y = x^{200} e^{-x}$. Eventually, and for surprisingly small values of x, the exponential term will always dominate. In fact, even in the last example the function starts to become overwhelmed by the exponential term around $x = 200$. However, for increasingly negative values of x, both terms in the product function are positive and increasing, thus ensuring that the product function increases more rapidly than either of its component terms. All of these features are apparent in Figure 2.19.

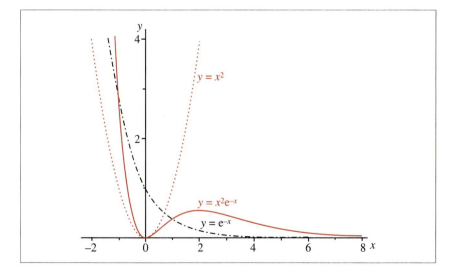

Figure 2.19 Plots of the functions $y = x^2$, $y = e^{-x}$ and the product $y = x^2 e^{-x}$

Product of a Polynomial or Trigonometric Function with an Exponential Function

The most common types of expression of this kind found in chemistry typically have one of the following forms:

- $P_n(x)e^{-x}$, where $P_n(x)$ is a **polynomial function** of degree n
- $\sin(nx)e^{-x}$

Polynomial functions have the general form

$$P_n(x) = c_0 + c_1 x + c_2 x^2 + c_3 x^3 + \cdots + c_n x^n \tag{2.43}$$

where $c_0, c_1,..., c_n$, are real constants and n is a positive integer, the largest value of which defines the degree of the polynomial. Polynomial functions of degree 3 or higher may display finite regions of oscillation (see Figure 2.20); in contrast, the trigonometric functions sin and cos oscillate indefinitely (see Figure 2.17).

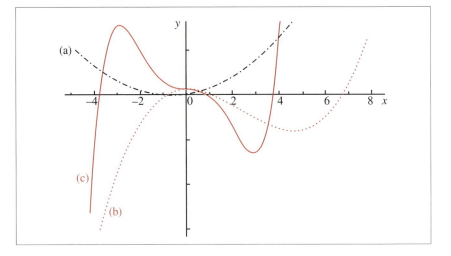

Figure 2.20 Plots of the polynomial functions (a) $y = 3x^2+4x+1$ (degree 2), (b) $y = x^3-7x^2 + x + 6$ (degree 3) and (c) $y = \frac{1}{2}x^5 - 7x^3 - x + 6$ (degree 5); the latter two display finite regions of oscillation

When we form a product of either a polynomial or a trigonometric function with the exponential function $y = e^{-x}$, the rapid decline in value of the exponential function as x increases from zero results in a rapid damping of the oscillation (see Figure 2.21).

In the case of polynomials of higher degree, a *finite* number of oscillations occur before they become overwhelmed by the exponential function, whereas for the product of sine or cosine with an exponential function the number of oscillations is *infinite*, with their amplitude decreasing with increasing positive x. For negative values of x the opposite occurs, with the amplitude of the oscillations increasing with increasingly negative values of x.

A Chemical Example: the 3s Atomic Radial Wave Function for the Hydrogen Atom

The radial part of the 3s atomic orbital function for the hydrogen atom is a good chemical example of a product of a polynomial with an exponential function, and takes the form:

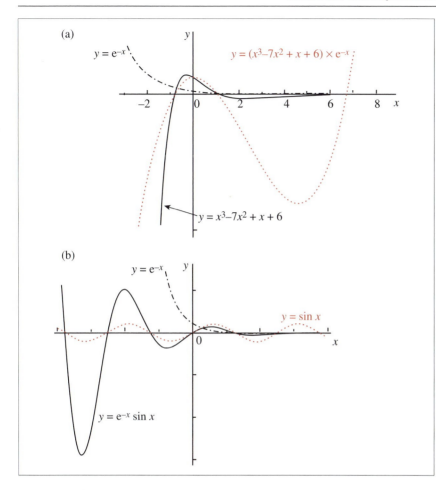

(a)

$y = e^{-x}$

$y = (x^3 - 7x^2 + x + 6) \times e^{-x}$

$y = x^3 - 7x^2 + x + 6$

(b)

$y = e^{-x}$

$y = \sin x$

$y = e^{-x} \sin x$

Figure 2.21 The product of (a) a polynomial or (b) a trigonometric function with the exponential function $y = e^{-x}$ results in a rapid damping of the oscillation as x increases

$$R_{3s} = P_2\left(\frac{r}{a_0}\right) e^{-r/3a_0} = N\left\{27 - \frac{18r}{a_0} + 2\frac{r^2}{a_0^2}\right\} e^{-r/3a_0} \qquad (2.44)$$

where N is a constant, having the form $\frac{1}{81a_0^{3/2}\sqrt{3\pi}}$. This may look rather complicated but it is in fact relatively straightforward in its form, comprising a second-degree polynomial function and an exponential function, both of which are expressed as a function of the independent variable r (the distance of the electron from the nucleus). In fact, wherever r appears in both parts of the product, it is divided by a_0, the Bohr radius, and we say in the case of the polynomial function that it is second degree in r/a_0 (the independent variable). A plot of $R_{3s}a_0^{3/2}$ versus r/a_0 is given in Figure 2.22. If we compare the plot with the function displayed in Figure 2.19 it should be clear that they both have essentially the same form. The main difference in the case of the radial wave function is that we consider only values of $r \geq 0$, simply because negative values for the radial distance have no physical meaning.

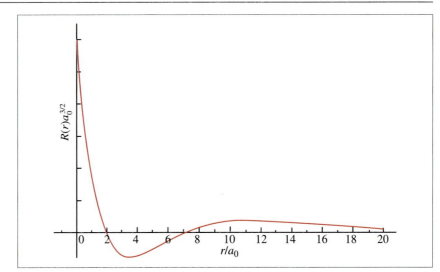

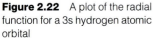

Figure 2.22 A plot of the radial function for a 3s hydrogen atomic orbital

2.3.5 Explicit and Implicit Functions

Up to this point, we have met functions of the form $y = f(x)$ in which the independent variable appears on the right side and the dependent variable appears on the left. In such cases, the association between a given value of the independent variable and the value of the dependent variable is explicit. For example, the function:

$$y = e^x$$

is an example of one in which y is an **explicit function** of x. However, we can always express such functions in the form $f(x,y) = 0$ in which y is an **implicit function** of x. For example, the function $y = e^x$ may be presented in an implicit form as:

$$\ln y - x = 0$$

In this example the implicit form of the function may be rearranged into a form in which either variable is an explicit function of the other. However, sometimes we meet functions which are impossible to arrange into an explicit form. The function:

$$y + e^y = x^5$$

is an example of an implicit function for which there is one unique value of y associated with each value of x but which cannot be expressed in the form of an explicit relationship between y and x. It is nevertheless possible in this case to compute each value of y associated with a particular value of x using numerical methods. An example from chemistry is the van der Waals equation (2.6) in which both P and T can be expressed as an explicit function of the other:

$$P = \frac{RT}{V_{\rm m} - b} - \frac{a}{V_{\rm m}^2}; \; T = \frac{(V_{\rm m} - b)}{R}\left(P + \frac{a}{V_{\rm m}^2}\right)$$

However, it is most convenient to consider the molar volume $V_{\rm m}$ as an *implicit function* of P and T (see equation 2.53).

2.4 Equations

Consider the plots of the quadratic polynomial functions $y = x^2 - 4x + 3$, $y = x^2 - 4x + 4$ and $y = x^2 - 4x + 6$ in Figure 2.23. Curve (a) cuts the x-axis ($y = 0$) at $x = 3$ and $x = 1$, values which correspond to the two solutions (or **roots**) of the **quadratic equation** $x^2 - 4x + 3 = 0$. In this example, we can more easily obtain the two roots by factorizing the **polynomial equation**, rather than by plotting the function. Thus $x^2 - 4x + 3$ can be expressed as the product of two linear factors:

$$x^2 - 4x + 3 = (x - 3)(x - 1)$$

and we can see that this will equal zero when either of the two linear factors equals zero, *i.e.*:

$$\text{when } x - 3 = 0 \Rightarrow x = 3$$
$$\text{or when } x - 1 = 0 \Rightarrow x = 1$$

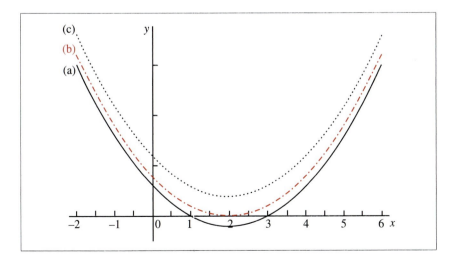

Figure 2.23 Plots of quadratic polynomial functions
(a) $y = x^2 - 4x + 3$,
(b) $y = x^2 - 4x + 4$,
(c) $y = x^2 - 4x + 6$

In cases where factorization proves difficult, it is always possible to use the formula for the roots of a quadratic equation, $ax^2 + bx + c = 0$:

$$x = \frac{-b \pm \sqrt{b^2 - 4ac}}{2a} \qquad (2.45)$$

In this example the coefficients a, b and c have values equal to 1, -4 and 3 and substituting these into our formula gives:

$$x = \frac{4 \pm \sqrt{16 - (4 \times 3)}}{2} = 2 \pm \frac{\sqrt{4}}{2} = 2 \pm 1$$

The quantity $b^2 - 4ac$ is known as the **discriminant**, and its value can be positive, zero or negative. In cases where it is positive, the equation has two real and different roots; if it is zero, then the equation will have two identical roots; and if it is negative, then there are no real roots, as the formula involves the square root of a negative number, for which there is no real result. A way around this latter difficulty is described in Chapter 2 of Volume 2, where **complex numbers** are introduced.

The value of the discriminant for the equation $x^2 - 4x + 3 = 0$ is positive, and we see that there are clearly two different roots, as indicated in plot (a) of Figure 2.23, which shows the curve cutting the x-axis at $x = 1$ and $x = 3$. The curve of the function $y = x^2 - 4x + 4$, shown in plot (b), touches the x-axis at $x = 2$. In this case the discriminant is zero, and we have two equal roots, given by $x = \frac{4}{2} \pm \sqrt{0} = 2 \pm 0$. Note that although the curve only touches the x-axis in one place, the equation $x^2 - 4x + 4 = 0$ still has two roots: they just happen to be identical. Finally, in the case of curve (c), there are no values of x corresponding to $y = 0$, indicating that there are no real roots of the quadratic equation $x^2 - 4x + 6 = 0$, as the discriminant is equal to -8.

Problem 2.11

(a) Use equation (2.45) to determine the number of real roots of the quadratic equations $f(x) = 0$, where $f(x)$ is given by: (i) $x^2 + x - 6$; (ii) $x^2 - 1$; (iii) $x^2 - 2\sqrt{2}x + 2$. (b) Give the factored form of each polynomial function $f(x)$.

In general, a quadratic equation has either two or zero real roots. However, a **cubic equation** may have one or three real roots, as seen in Worked Problem 2.5.

Worked Problem 2.5

Q Use a graphical method to find the number of roots of the polynomial equations: (a) $x^3 - 7x + 6 = 0$; (b) $x^3 - 4x^2 - 2x - 3 = 0$.

A (a) 3; (b) 1 (see Figure 2.24).

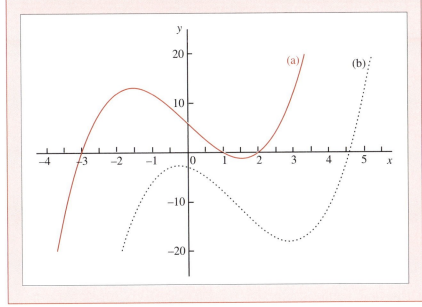

Figure 2.24 Plots of the polynomial functions (a) $x^3 - 7x + 6$, (b) $x^3 - 4x^2 - 2x - 3$

2.4.1 An Algebraic Method for Finding Roots of Polynomial Equations

For a given polynomial function $y = f(x)$, one (or more) roots of the polynomial equation $f(x) = 0$ can often be found by an algebraic method. Suppose the polynomial $f(x)$ is of degree n. If $x = \lambda$ is a root of the polynomial equation, then $f(\lambda) = 0$, and $(x - \lambda)$ is a factor of the polynomial:

$$f(x) = (x - \lambda)(c_1 x^{n-1} + c_2 x^{n-2} + \dots + c_n) \tag{2.46}$$

The truth of the previous statement follows by substituting $x = \lambda$ into the above equation, where we see that, irrespective of the value of the second expression in parentheses, which is a polynomial of degree $n-1$, the first term in parentheses is zero, thus implying that $f(\lambda) = 0$. If there is a root with integer value, then it can sometimes be found by trial and error, using $\lambda = \pm 1, \pm 2, \dots$ the polynomial of degree $n-1$ can then be treated in the same way. If no further roots can be found algebraically, at any stage in the iterative procedure, then the current polynomial can be plotted to exhibit the existence, or otherwise, of remaining roots.

The key requirement is that, at each step, the coefficients c_i are found, in order to facilitate the recovery of another root. Once the polynomial of degree two is reached, it is easiest to use the formula given in equation (2.45) to test for the existence of a further two or zero roots.

Worked Problem 2.6

Q   (a) Use the algebraic method to find the roots of the polynomial equation $f(x) = 0$, where $f(x) = 2x^3 + 11x^2 + 17x + 6 = 0$; (b) give the factored form of $f(x)$; (c) sketch a graph of the function $y = f(x)$.

A   (a) Simple trial and error shows that $x = -2$ is a root of $f(x)$, since $f(-2) = 0$. The polynomial equation may now be written in the form:

$$(x + 2)(c_1 x^2 + c_2 x + c_3) = 0$$

On multiplying out the brackets, and collecting terms, we have:

$$c_1 x^3 + (c_2 + 2c_1)x^2 + (c_3 + 2c_2)x + 2c_3 = 0$$

Comparing coefficients of the powers of x with the given polynomial equation, we find:

$$c_1 = 2 \tag{2.47}$$

$$c_2 + 2c_1 = 11 \tag{2.48}$$

$$c_3 + 2c_2 = 17 \tag{2.49}$$

$$2c_3 = 6 \tag{2.50}$$

Equations (2.47) and (2.50) give the values $c_1 = 2$ and $c_3 = 3$, respectively. It then follows, by substituting the value of c_1 in equation (2.48), that $c_2 = 7$, and we then have:

$$(x + 2)(2x^2 + 7x + 3) = 0$$

(b) The solutions of $(2x^2 + 7x + 3) = 0$ are then found using equation (2.45):

$$x = -\frac{7}{4} \pm \frac{1}{4}\sqrt{49-24} = -\frac{7}{4} \pm \frac{5}{4} = -3 \text{ or } -\frac{1}{2}$$

Thus,

$$f(x) = (x+2)\left(x+\frac{1}{2}\right)(x+3).$$

(c) From (b) we know that the curve crosses the x-axis at $x = -2, -3,$ $-1/2$; in addition, for $x > -1/2$, all three brackets are positive and increase in value as x increases. Likewise, for $x < -3$, all brackets have increasing negative values, and therefore $f(x)$ is negative for these values. For $-3 < x < -2$, $(x+3)$ is positive and $(x+2)$ and $(x+1/2)$ are both negative, and hence $f(x) > 0$. A similar argument shows that $f(x) < 0$ for $-2 < x < -1/2$, and it is then an easy matter to sketch the form of the cubic polynomial function (Figure 2.25).

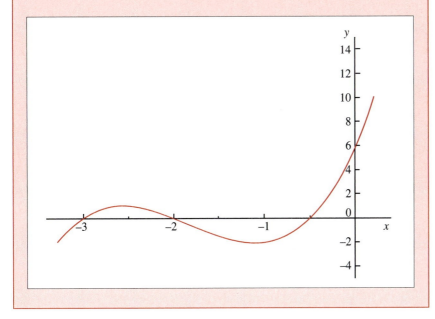

Figure 2.25 Plot of the function $f(x) = 2x^3 + 11x^2 + 17x + 6 = 0$

2.4.2 Solving Polynomial Equations in a Chemical Context

In practice, the solution of polynomial equations is problematic if no simple roots are found by trial and error. In such circumstances the graphical method may be used or, in the cases of a quadratic or cubic equation, there exist algebraic formulae for determining the roots. Alternatively, computer algebra software (such as Maple or Mathematica, for example) can be used to solve such equations explicitly. In Worked Problem 2.7 we show how the calculation of the pH of 10^{-6} mol dm^{-3} HCl(aq) requires the solution of a quadratic equation.

Worked Problem 2.7

Q Calculate the pH of 10^{-6} mol dm^{-3} HCl(aq), taking into account the hydronium (H$_3$O$^+$) ions from: (a) HCl alone; (b) HCl and the dissociation of water (equilibrium constant, $K_w = 10^{-14}$).

A (a) The simple formula pH $= -\log([\text{H}_3\text{O}^+]/\text{mol dm}^{-3})$ leads to a value for the pH of 6, since $[\text{H}_3\text{O}^+] = 10^{-6}$ mol dm^{-3}. (b) As the concentration of HCl is so small, it is appropriate to take account of the dissociation of water in our calculation of the pH, and so we need to consider the concentration of hydronium ions produced from two sources, described by the following processes:

$$\text{HCl} + \text{H}_2\text{O} \longrightarrow \text{H}_3\text{O}^+ + \text{Cl}^-$$

$$2\text{H}_2\text{O} \underset{}{\overset{K_w}{\rightleftharpoons}} \text{H}_3\text{O}^+ + \text{OH}^-$$

Thus, if $[\text{H}_3\text{O}^+]/\text{mol dm}^{-3} = h$, $[\text{Cl}^-]/\text{mol dm}^{-3} = c$ and $[\text{OH}^-]/\text{mol dm}^{-3} = b$, then charge conservation requires:

$$h = c + b \Rightarrow b = h - c$$

where $c = 10^{-6}$. The equilibrium constant for the dissociation of water is given by $K_w = hb$, which we can now rewrite as:

$$K_w = hb = h(h - c) = h^2 - ch \Rightarrow h^2 - ch - K_w = 0 \qquad (2.51)$$

Equation (2.51) is a quadratic equation in h, and the two roots may be found using equation (2.45). Thus:

$$h = \frac{c}{2} \pm \frac{\sqrt{c^2 + 4K_w}}{2}$$

and, on substituting for c and K_w, we find $h = 1.099 \times 10^{-6}$ or $h = -9.902 \times 10^{-9}$. The first solution yields pH $= 5.996$; the second solution, although mathematically required, does not correspond to a acceptable physical result, as the logarithm of a negative number is not defined as a real number and thus has no physical significance.

Charge conservation requires that there are the same number of cations as anions in the solution. Thus the sum of the concentrations of the OH$^-$ and Cl$^-$ ions must be the same as the hydronium ion concentration and so $h = c + b$.

The logarithm of a negative number is a so-called complex number which we discuss in some detail in Chapter 1 of Volume 2.

Problem 2.12

The radial function of the 3s atomic orbital for the hydrogen atom has the form given in equation (2.44).

(a) Calculate the value of R_{3s} at $r=0$ and as r tends to infinity. Note that the exponential term will always dominate the term in parentheses (see Section 2.3.4) and so its limiting behaviour alone will determine the behaviour of the function as r tends to infinity (see also Chapter 3 for a more detailed discussion of limits).

(b) Calculate the values of r/a_0, and hence of r, for which $R_{3s}=0$, by solving the quadratic polynomial equation $\{27 - 18(\frac{r}{a_0})+2(\frac{r}{a_0})^2\} = 0$.

(c) Sketch the form of R_{3s} for $0 \leqslant \frac{r}{a_0} \leqslant 12$, and then compare your result with that displayed in Figure 2.22.

Polynomial Equations of Higher Degree in Chemistry

Polynomial equations of degree three (cubic equations) arise in a number of areas of classical physical chemistry; however, such equations also arise in the modelling of:

- Electronic structures, through the determination of molecular orbitals, constructed as linear combinations of atomic orbitals (LCAO); thus, for example, the determination of the simplest σ-type molecular orbitals for HCN, in its linear configuration (as in the ground state), involves the use of the seven σ atomic orbitals $1s_H$, $1s_C$, $1s_N$, $2s_C$, $2s_N$, $2p\sigma_C$ and $2p\sigma_N$, and leads to the solution of a polynomial equation of degree seven for the molecular orbital energies.
- Characteristic frequencies of molecular vibrations. In the case of HCN, for example, there are four vibrational frequencies that may be calculated from a polynomial equation of degree four, by making appropriate assumptions about the stiffness of bond stretching and bond angle deformation.

Problem 2.13

Give the degree of the polynomial equation that arises in calculating the molecular orbitals for the following species in their ground states (σ or π bonding, as indicated): (a) carbon dioxide (σ only); (b) benzene (π bonds only).

Examples of a Cubic Polynomial Equation in Physical Chemistry

The van der Waals Equation Revisited

Consider the relationship between pressure, temperature and volume that is provided by the van der Waals equation, used to model the physical properties of a real, rather than an ideal, gas:

$$P = \frac{RT}{V_m - b} - \frac{a}{V_m^2} \tag{2.52}$$

Here, a and b are parameters for a specific gas, and V_m is the molar volume. If we now multiply both sides of this equation by $V_m^2(V_m - b)$, and rearrange the terms, the following cubic equation results:

$$V_m^3 - V_m^2\left(b + \frac{RT}{P}\right) + V_m\frac{a}{P} - \frac{ab}{P} = 0 \tag{2.53}$$

We can use this third-degree polynomial to find the molar volume of a gas at a given temperature, T, and pressure, p. For example,[3] we can estimate the molar volume of CO_2 at 500 K and 100 atm using the literature values[2] for $a = 3.592$ atm dm^6 mol^{-1} and $b = 0.04267$ dm^6 mol^{-1} and taking $R = 0.082058$ atm dm^6 K^{-1} mol^{-1}. The solution to equation (2.53) yields only one real root (the other two roots are complex), and we obtain a value $V_m = 0.3663$ dm^3 (found using Maple, the computer algebra software). The plot of the function (Figure 2.26) confirms this finding.

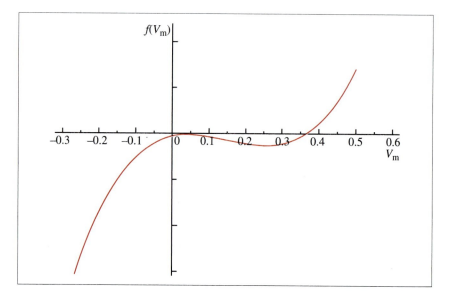

Figure 2.26 Plot of the van der Waals polynomial function for CO_2 using $P = 100$ atm and $T = 500$ K

The 4s Radial Wavefunction for the Hydrogen Atom

In order to locate the **nodes** in the radial part of the hydrogen 4s atomic orbital:

A node is a point where the wavefunction passes through zero.

$$R_{4s} = N \left\{ 24 - 18\left(\frac{r}{a_0}\right) + 3\left(\frac{r}{a_0}\right)^2 - \frac{1}{8}\left(\frac{r}{a_0}\right)^3 \right\} \tag{2.54}$$

we need to solve a cubic polynomial equation for the three values of $\frac{r}{a_0}$, and hence r, as a multiple of a_0. Unlike the simple expressions for the solutions of a quadratic equation given in equation (2.45), and the cubic equation in Worked Problem 2.6, a more involved algebraic procedure is required to solve the cubic equation given in equation (2.54). However, we know that a hydrogen 4s atomic orbital has three radial nodes $(n-l-1)$, and since there are three roots to the third-order polynomial equation $R_{nl}(r) = 0$, we conclude that all three roots are real. In this case, therefore, the graphical method will give good estimates for the location of the roots, which can then be improved by trial and error; alternatively, computer algebra software can be used to determine the roots to any sensible number of decimal places.

Problem 2.14[4]

When 1.00 mol of sodium ethanoate (NaOAc) is dissolved in 1.00 dm^3 of water, Na$^+$ ions are liberated and some of the AcO$^-$ species combine with H$_3$O$^+$ to form ethanoic acid (AcOH). The following equilibria are set up:

$$\text{AcOH} + \text{H}_2\text{O} \overset{K}{\rightleftharpoons} \text{H}_3\text{O}^+ + \text{AcO}^- \quad \text{and} \quad 2\text{H}_2\text{O} \overset{K_w}{\rightleftharpoons} \text{H}_3\text{O}^+ + \text{OH}^-$$

where:

$$K = \frac{a_{\text{H}_3\text{O}^+} a_{\text{AcO}^-}}{a_{\text{AcOH}}} \tag{2.55}$$

and:

$$K_w = a_{\text{H}_3\text{O}^+} a_{\text{OH}^-} \tag{2.56}$$

If the activities a_{AcOH}, $a_{\text{H}_3\text{O}^+}$, a_{OH^-} and a_{AcO^-} are designated by the four unknowns e, h, b and c, then four equations need specifying in order to solve for the unknowns.

(a) Use equations (2.55) and (2.56) to give the defining relations for K and K_w in terms of e, h, b and c.

(b) Confirm that the conservation requirements for AcO$^-$ and charge (concentrations of cations and anions are equal) yield equations (2.57) and (2.58):

$$c + e = 1 \qquad (2.57)$$

$$c + b = 1 + h \qquad (2.58)$$

(c) Rearrange equation (2.57) to find an expression for e.

(d) Obtain an expression for c in terms of K and h, by substituting for e in equation (2.55).

(e) Substitute the expression for c into equation (2.58), and show that $b = 1 + h - \frac{K}{K+h}$.

(f) Substitute for b in equation (2.56), and show that $K_w = h + h^2 - \frac{Kh}{K+h}$.

(g) Multiply the first two terms on the right side of the result in (f) by $\frac{K+h}{K+h}$, and rearrange the expression in (f) to show that the following cubic equation is obtained for determining h:

$$h^3 + h^2(1 + K) - K_w h - K_w K = 0 \qquad (2.59)$$

(h) Substitute the values $K = 1.8 \times 10^{-5}$ and $K_w = 10^{-14}$ into equation (2.59) to obtain a cubic equation with numerical coefficients.

(i) Given that equation (2.59) has $h = 4.243 \times 10^{-9}$ as its only physically meaningful root (the other two roots are negative and therefore meaningless), find the values for e, b and c and hence of a_{AcOH}, a_{OH^-} and a_{AcO^-} to three significant figures.

The exact definition of the equilibrium constant given by IUPAC requires it to be defined in terms of fugacity coefficients or activity coefficients, in which case it carries no units. This convention is widely used in popular physical chemistry texts, but it is also common to find the equilibrium constant specified in terms of molar concentrations, pressure or molality, in which cases the equilibrium constant will carry appropriate units.

Summary of Key Points

This chapter revolves around the important concepts of function, equation and formula. The key points discussed include:

1. Function as an association between one number and another; the domain, as a set which specifies the numbers for which the association applies.

2. The independent and dependent variables in an association and their identification with particular, yet arbitrary, symbols in a formula.

3. The role of units in working with functions in a chemical context; creating a function from a formula.

4. Representing functions in tabular, formula, prescription or graphical forms; how the choice of domain affects the appearance of a plot.

5. The definition of the domain for a chemical function as opposed to an abstract function.

6. Testing a chemical formula by comparison of real experimental data with that generated from a model.

7. Special mathematical functions: exponential, logarithm (base 10 and base e), trigonometric, reciprocal trigonometric and hyperbolic trigonometric.

8. Working with the properties of logarithms and trigonometric identities.

9. Measurement of angles: degrees and radians.

10. Symmetric, antisymmetric and periodic functions; product functions; the product of a polynomial function or a trigonometric function with an exponential function.

11. Equations; solving quadratic and higher order equations; finding the factors and roots of simple polynomial equations using either algebraic or graphical procedures.

References

1. Data taken from W. R. Moore, *Physical Chemistry*, 5th edn., Longman, Harlow, UK, 1976, p. 23.
2. Data from *Handbook of Chemistry and Physics*, 58th edn., CRC Press, Boca Raton, FL, USA, 1978, Table D178.
3. Example taken from P. W. Atkins and J. de Paula, *Physical Chemistry*, 7th edn., Oxford University Press, Oxford, 2002, p. 20.
4. Adapted from M. J. Sienko, *Equilibrium*, part 2, Benjamin, New York, 1964, p. 511.

3

Limits

The concept of the limit is a fairly broad one, commonly used for probing the behaviour of mathematical functions as the independent variable approaches a particular value, either in exploring errant or unexpected behaviour or in examining the behaviour of functions as the independent variable takes on increasingly large or small positive or negative values. More importantly, limits are central to our understanding of differential calculus, as is seen in the work of Fermat who, in the early 17th century, used the concept of the limit for finding the slope of the tangent at a point on a curve (a topic discussed in Chapter 4). Likewise, in Chapter 6 we shall see how the concept of the limit provides a foundation for integral calculus.

Aims

By the end of the chapter you should be able to:

- Understand the principles involved in defining the limit
- Understand the notions of continuity and discontinuity
- Use limits to examine the point behaviour of functions which might display unexpected characteristics
- Investigate asymptotic behaviour of functions as the independent variable takes on increasingly large positive or negative values
- Use limits to improve your understanding of how physical processes may change as experimental conditions change from one extreme to another

3.1 Mathematical and Chemical Examples

3.1.1 Point Discontinuities

The function shown in Figure 3.1 shows a break at $x = 3$, where the value of y is $\frac{0}{0}$, and is therefore indeterminate. In this situation the function is said to exhibit a **discontinuity** at $x = 3$, which means that it is impossible to sketch the plot of the function by hand without taking the pencil off the paper.

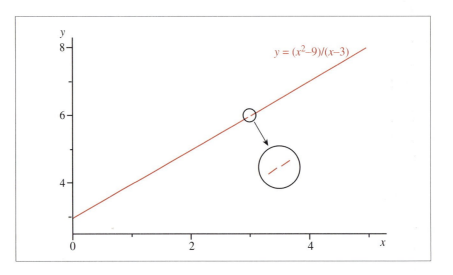

Figure 3.1 A plot of the function $y = (x^2 - 9)/(x - 3)$ over the subinterval $0 \le x \le 5$

A chemical example is shown schematically in Figure 2.8, where discontinuities are seen in the entropy function at the melting and boiling points, T_m and T_b, respectively, as well as at a temperature T_s where a change in crystal structure occurs in the solid state. Although the entropy function is undefined at these three transition temperatures, the discontinuities are finite in nature, as the corresponding changes in S are finite in size. We can see from this example that S is continuous only over sub-intervals of the domain; furthermore, at each of the transition temperatures, T_m, T_b and T_s, the value of S is ambiguous. This situation arises because two values of S result, depending on whether we approach a transition temperature from higher or lower values of T.

Sometimes we meet functions displaying an **infinite discontinuity**. For example, the function $y = f(x) = 1/(1 - x)$, shown in Figure 3.2, displays such a discontinuity at $x = 1$ because as we approach $x = 1$ from higher and lower values of x the value of $f(x)$ tends towards infinitely large values in negative and positive senses, respectively. In this example the line $x = 1$ is known as a vertical asymptote (see Sections 2.3.1 and 2.3.3 for further discussion of asymptotic behaviour).

The tangent function, $\tan x = \sin x / \cos x$, shown in Figure 3.3, is interesting because it exhibits infinite discontinuities whenever x passes through an odd multiple of $\frac{\pi}{2}$.

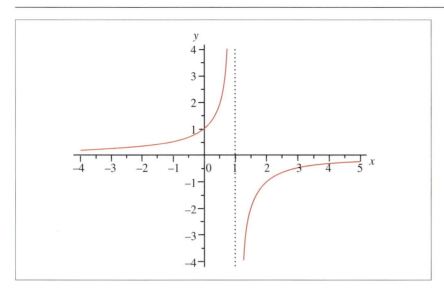

Figure 3.2 A plot of the function $f(x) = 1/(1 - x)$

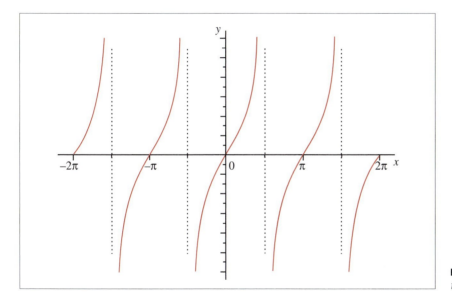

Figure 3.3 A plot of the function $f(x) = \tan x$

3.1.2 Limiting Behaviour for Increasingly Large Positive or Negative Values of the Independent Variable

We now turn to examining the **limiting behaviour** of functions as the independent variable takes on increasingly large positive or negative values. As an illustration, consider the function shown in Figure 3.2. We see from the form of $f(x)$ that the value of y approaches zero as x becomes increasingly large in both positive and negative senses: the line $y = 0$ is an **asymptote**. In the former case the values of y are increasingly small negative numbers and in the latter they are increasingly small positive numbers. The **limiting values** of $f(x)$ are therefore zero in both cases.

Periodic functions such as $\sin x$ or $\cos x$ have no asymptotes (no single limiting value), because their values oscillate between two limits as the independent variable increases in a positive or negative sense. For example, the value of the function $f(x) = \cos(2x)$ oscillates between $+1$ and -1 as $x \rightarrow \infty$ (see Figure 3.4).

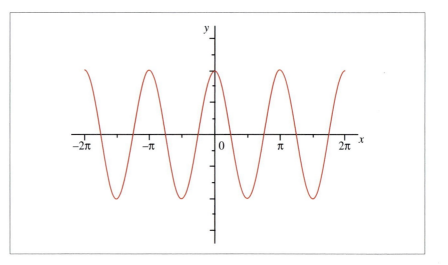

Figure 3.4   A plot of the function $f(x) = \cos(2x)$

Problem 3.1

Find the limiting values for (a) $x^2 e^{-x}$ and (b) $\cos(2x)e^{-x}$ as $x \rightarrow \infty$.

3.1.3 Limiting Behaviour for Increasingly Small Values of the Independent Variable

Frequently, the context of a particular problem requires us to consider the limiting behaviour of a function as the value of the independent variable approaches zero. For example, consider the physical measurement of heat capacity at absolute zero. Since it is impossible to achieve absolute zero in the laboratory, a natural way to approach the problem would be to obtain measurements of the property at increasingly lower temperatures. If, as the temperature is reduced, the corresponding measurements approach some value m, then it may be assumed that the measurement of the property (in this case, heat capacity) at absolute zero is also m, so long as the specific heat function is continuous in the region of study. We say in this case that the limiting value of the heat capacity, as the temperature approaches absolute zero, is m. As we shall see in Section 3.2, the notation we use to describe this behaviour is:

$$\lim_{T \to 0} C_V(T) = m \qquad (3.1)$$

where, in this case, $m = 0$ because the limiting value of the heat capacity as $T \to 0$ K is zero. It is also important to note that it is only possible to approach absolute zero from positive values of T; thus, in this situation, the "right" limit, usually written as $\lim_{T \to 0^+} C_V(T) = m$, is the only one of physical significance.

Problem 3.2

Find the limiting values for (a) $x^2 e^{-x}$ and (b) $\cos(2x)e^{-x}$ as $x \to 0$.

3.2 Defining the Limiting Process

For a function of a single variable x, symbolized, as usual, by $y = f(x)$, we are interested in the value of $f(x)$ as x approaches a particular value, a, but never takes the value a. Points where the function is not defined, as seen, for example, at $x = 1$ in Figure 3.2, are excluded from the domain of the function; at other points, the function is continuous.

  Limits play an important role in probing the behaviour of a function at any point in its domain, and the notation we use to describe this process is:

$$\lim_{x \to a} f(x) = m \qquad (3.2)$$

Note: in this symbolism, the suffix to the symbol lim indicates that, although x approaches a, it *never* actually takes the value a. For the limit to exist, the same (finite) result must be obtained whether we approach a from smaller or larger values of x. Furthermore, if $m = f(a)$, then the function is said to be continuous at $x = a$.

3.2.1 Finding the Limit Intuitively

Consider the plot of the function:

$$y = f(x), \text{ where } f(x) = \frac{x^2 - 9}{x - 3} \qquad (3.3)$$

shown in Figure 3.1. It is evident that $f(x)$ is continuous (unbroken) for all values of x except $x = 3$. Since the denominator and numerator of the function are both zero at $x = 3$, we see that the function is indeterminate at this value of x; however, as seen in Table 3.1, the *ratio* of the numerator and denominator seems to be approaching the value $y = 6$ as $x \to 3$ from smaller or larger values.

Table 3.1   Values of $f(x) = (x^2-9)/(x-3)$ in the vicinity of $x = 3$

x	$x^2 - 9$	$x - 3$	$(x^2 - 9)/(x - 3)$
4	7	1	7
3.5	3.25	0.5	6.5
3.1	0.61	0.1	6.1
3.01	0.0601	0.01	6.01
3	0	0	indeterminate
2.99	-0.0599	-0.01	5.99
2.9	-0.59	-0.1	5.9
2.5	-2.75	-0.5	5.5
2	-5	-1	5

Taking even smaller increments either side of 3, say $x = 3 \pm 0.0001$, we find that $f(3.0001) = 6.0001$ and $f(2.9999) = 5.9999$. These results suggest that for smaller and smaller increments in x, either side of $x = 3$, the values of the function become closer and closer to 6. Thus we say that, in the limit as $x \to 3$, m takes the value 6:

$$\lim_{x \to 3} \frac{x^2 - 9}{x - 3} = 6 \tag{3.4}$$

3.2.2 An Algebraic Method for Evaluating Limits

In practice, it is often easiest when evaluating limits to write $x = a \pm \delta$, and consider what happens as $\delta \to 0$, *but never takes the value zero*. This procedure allows us to let x become as close as we like to the value a, *without* it taking the value $x = a$.

Worked Problem 3.1

Q   Evaluate $\lim_{x \to 3} f(x)$, where $f(x) = \dfrac{x^2 - 9}{x - 3}$.

A   By substituting $x = 3 + \delta$ in the expression for $f(x)$, and expanding the square term in the numerator, we obtain:

$$\lim_{x \to 3} \frac{x^2 - 9}{x - 3} = \lim_{\delta \to 0} \frac{(3 + \delta)^2 - 9}{3 + \delta - 3}$$

$$= \lim_{\delta \to 0} \frac{9 + 6\delta + \delta^2 - 9}{\delta} = \lim_{\delta \to 0} \frac{6\delta + \delta^2}{\delta}$$

$$= \frac{6 - \delta}{1} = 6$$

where, in the last step, δ can be cancelled in every term of the numerator and denominator as its value is never zero. Thus we obtain the expected result that $f(x)$ approaches the limiting value of 6 as x tends to the value 3, irrespective of the sign of δ. In this situation, m in the definition of the limit has the value 6.

Problem 3.3

For each of the following functions, $f(x)$, identify any points of discontinuity (those values of x where the function is of indeterminate value) and use the method described in Worked Problem 3.1, where appropriate, to find the limiting values of the following functions at your suggested points of discontinuity.

(a) $f(x) = \dfrac{2x}{x - 4}$; (b) $f(x) = \dfrac{x^2 - 4}{x - 2}$; (c) $f(x) = \dfrac{x - 1}{x^2 - 1}$;

(d) $f(x) = 3x^2 - \dfrac{2}{x} - 1$.

3.2.3 Evaluating Limits for Functions whose Values become Indeterminate

Whenever the value of a function becomes indeterminate for particular limiting values in the independent variable (for example, division by zero or expressions such as ∞/∞ or $\infty - \infty$), we need to adopt alternative strategies in determining the limiting behaviour. Such situations arise quite commonly in chemistry, especially when we are interested in evaluating some quantity as the independent variable takes on increasingly large or small values. Good examples occur in dealing with mathematical expressions arising in:

* Manipulating the solutions of rate equations in kinetics.
* Determining high- or low-temperature limits of thermodynamic properties.

Worked Problem 3.2

Q Find $\lim\limits_{x\to\infty} \dfrac{2x^2+4}{x^2-x+1}$.

A Both the numerator and denominator tend to infinity as $x\to\infty$, but their ratio remains finite. There are two ways of handling this situation:

First, we note that as x becomes very large, $2x^2+4$ is increasingly well approximated by $2x^2$, and x^2-x+1 by x^2 as, in both expressions, the highest power of x dominates as x becomes indefinitely large. Thus, as x increases without limit, we find:

$$\lim_{x\to\infty}\frac{2x^2+4}{x^2-x+1}=\lim_{x\to\infty}\frac{2x^2}{x^2}=\lim_{x\to\infty}2=2$$

Second, we could divide the numerator and denominator by the highest power of x, before taking the limit:

$$\lim_{x\to\infty}\frac{2x^2+4}{x^2-x+1}=\lim_{x\to\infty}\frac{2+4/x^2}{1-1/x+1/x^2}=2$$

and, again, we see that as x increases without limit, the ratio of numerator to denominator tends to 2.

Problem 3.4

Evaluate the following limits:

(a) $\lim\limits_{x\to\infty}\dfrac{5}{x+1}$; (b) $\lim\limits_{x\to\infty}\dfrac{3x}{x-4}$; (c) $\lim\limits_{x\to\infty}\dfrac{x^2}{x+1}$; (d) $\lim\limits_{x\to\infty}\dfrac{x+1}{x+2}$.

The limiting behaviour of functions for increasingly small values of the independent variable can be found in a similar way by applying exactly the same principles, except that, now, the lowest power of x provides the largest term in both numerator and denominator.

Worked Problem 3.3

Q Find $\lim\limits_{x\to 0}\dfrac{x^2+x}{x^3-1}$.

A This time, for increasingly small values of x, the numerator and denominator are dominated by x and -1, respectively. Consequently, the ratio of the numerator to denominator tends to $\frac{x}{-1}$, which leads to a limiting value of zero: $\lim\limits_{x\to 0}\frac{x}{-1}=0$.

Problem 3.5

Evaluate the limit $\lim\limits_{x \to 0}(\ln x - \ln 2x)$.
Hint: remember that $\ln a - \ln b = \ln\frac{a}{b}$ (see Chapter 2).

Problem 3.6

The Einstein model for the molar heat capacity of a solid at constant volume, C_V, yields the formula:

$$C_V = 3R(ax)^2 \left\{ \frac{e^{ax/2}}{e^{ax} - 1} \right\}^2$$

where $a = \frac{h\nu}{k}$ and $x = \frac{1}{T}$. Find the limiting value of C_V as $T \to 0$ K, remembering that $x = \frac{1}{T}$.

Note: we shall revisit this problem in Chapter 1 of Volume 2, where we explore the limiting behaviour for high values of T (Problem 1.10, Volume 2).

Problem 3.7

The radial function for the 3s atomic orbital of the hydrogen atom has the form:

$$R_{3s} = N\left(\frac{r}{a_0}\right)^2 e^{-r/a_0}$$

where N is a constant. Find the values of R_{3s} as: (a) $r \to 0$; (b) $r \to \infty$.
Hint: see your answers to Problems 3.1(a) and 3.2(a).

3.2.4 The Limiting Form of Functions of More Than One Variable

Sometimes, we are interested in how the form of a function might change for limiting values in one or more variables. For example, consider the catalytic conversion of sucrose to fructose and glucose by the enzyme invertase (β-fructofuranidase). The rate of formation of product P for this reaction varies in a rather complicated way with the sucrose concentration [S]. At low [S], the reaction is first order in [S], and at high [S] it is zero order. The behaviour observed in Figure 3.5 is established by investigating the form of the function describing the rate of reaction for the two limiting cases where [S] approaches either very large or very small values, rather than the absolute value of the function as in the examples discussed above. This is a consequence in this case of the rate equation being a function of more than one variable.

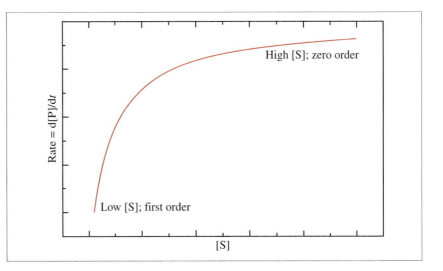

Figure 3.5 The variation in rate of enzymolysis for low and high sucrose concentration, [S], where the reaction is first and zero order, respectively

Worked Problem 3.4

Q The rate of formation of the product P in the catalytic conversion of sucrose to fructose and glucose by the enzyme invertase is given by:

$$\frac{d[P]}{dt} = \frac{k_2[E]_0[S]}{K_M + [S]}$$

where k_2 is a rate constant, K_M is known as the Michaelis constant, $[E]_0$ is the initial enzyme concentration and [S] is the sucrose concentration. Find the order of reaction with respect to [S] when (a) $[S] \gg K_M$ and (b) $[S] \ll K_M$.

A (a) For $[S] \gg K_M$, $K_M + [S] \approx [S]$ and so $\frac{d[P]}{dt} = \frac{k_2[E]_0[S]}{K_M + [S]}$ $\approx \frac{k_2[E]_0[S]}{[S]} = k_2[E]_0$: zero order in [S].

(b) For $[S] \ll K_M$, $K_M + [S] \approx K_M$ and $\frac{d[P]}{dt} = \frac{k_2[E]_0[S]}{K_M + [S]} \approx \frac{k_2[E]_0[S]}{K_M}$: first order in [S].

Problem 3.8

A rate law derived from a steady-state analysis of a reaction mechanism proposed for the reaction of H_2 with NO is given by:

$$\frac{d[N_2]}{dt} = \frac{k_1 k_2 [H_2][NO]^2}{k_{-1} + k_2[H_2]}$$

Find the limiting form of the rate law when (a) $k_{-1} \gg k_2[H_2]$ and (b) $k_{-1} \ll k_2[H_2]$.

Summary of Key Points

This chapter introduces the concept of the limit, with a view not only to probing limiting behaviour of functions but also as a foundation to the development of differential and integral calculus in the following chapters. The key points discussed include:

1. The principles involved and notation used in defining a limit.

2. Point discontinuities, infinite discontinuities and asymptotic behaviour.

3. Finding a limit intuitively and algebraically.

4. Investigating the limiting value of functions for increasingly large and small values of the independent variable.

5. Finding the limiting forms of functions of more than one variable.

4
Differentiation

A great deal of chemistry is concerned with processes in which properties change as a function of some variable. Good examples are found in the field of chemical kinetics, which is concerned with measuring and interpreting changes in concentrations of reactants or products with time, and in quantum mechanics, where we are interested in the rate of change in the electronic wavefunction of a diatomic molecule as a function of bond length.

Aims

Calculus is of fundamental importance in chemistry because it underpins so many key chemical concepts. In this chapter, we discuss the foundations and applications of **differential calculus**; by the end of the chapter you should be able to:

- Describe processes involving change in one independent variable
- Define the average rate of change of the dependent variable
- Use the concepts of limits to define the instantaneous rate of change
- Differentiate most of the standard mathematical functions by rule
- Differentiate a sum, product or quotient of functions
- Apply the chain rule to non-standard functions
- Understand the significance of higher-order derivatives and identify maxima, minima and points of inflection
- Understand the concept of the differential operator
- Understand the basis of the eigenvalue problem and identify eigenfunctions, eigenvalues and operators
- Differentiate functions of more than one variable

4.1   The Average Rate of Change

Consider the plot of the function $y = f(x)$, in which x is the independent variable, shown in Figure 4.1. The **average rate of change** of $f(x)$ over the increment Δx in x is given by:

$$\frac{QR}{PR} = \frac{f(x_0 + \Delta x) - f(x_0)}{\Delta x} = \frac{\Delta y}{\Delta x} \tag{4.1}$$

where $f(x_0)$ and $f(x_0 + \Delta x)$ are the values of $f(x)$ at the points x_0 and $x_0 + \Delta x$, and Δy is the change in y that results in the change Δx in x.

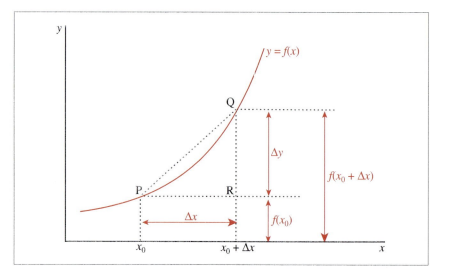

Figure 4.1   Defining the average rate of change of $f(x)$ as x is incremented from x_0 to $x_0 + \Delta x$

This average rate of change corresponds to the slope of the **chord** PQ; that is, the slope of the straight line (sometimes termed the **secant**) joining P and Q. In chemical kinetics, we can draw a direct analogy by equating the concentration of a species A at time t, often designated by [A], to the dependent variable (designated as y in Figure 4.1), and the time after initiation of the reaction, t, to the independent variable (designated by x in Figure 4.1). Consequently, if we measure the concentration of a reaction product at two intervals of time, say one minute apart, we might conclude that over that interval the concentration of the product had changed by 1.00 mol dm^{-3}. In this case, we could state that the average rate of reaction in this interval is 1.00 mol dm^{-3} per minute. The problem here is that we know nothing about how the reaction rate changes in detail during that interval of one minute, and it is this detail that is so crucial to our understanding of the kinetics of the reaction. Consequently, what we need, in general, is to be able to quantify the rate of change of the dependent variable at a *particular* value of the independent variable rather than simply the average rate of change over some increment in the independent variable. This equates, in our chemical analogy, to being able to measure the instantaneous reaction rate at a given instant in time (and consequently for a given concentration of reactant or product), rather than the average rate of reaction over some extended period of time. However, before we can determine these instantaneous chemical rates, we must first establish some mathematical principles.

4.2 The Instantaneous Rate of Change

4.2.1 Differentiation from First Principles

If we now reconsider the general situation shown in Figure 4.1, we can determine the **instantaneous rate of change** by examining the limiting behaviour of the ratio, QR/PR, the change in y divided by the change in x, as Δx tends to zero:

$$\lim_{\Delta x \to 0} \left\{ \frac{QR}{PR} \right\} = \lim_{\Delta x \to 0} \left\{ \frac{\Delta y}{\Delta x} \right\} = \lim_{\Delta x \to 0} \left\{ \frac{f(x_0 + \Delta x) - f(x_0)}{\Delta x} \right\} \qquad (4.2)$$

The limiting value defined in equation (4.2) exists if:

- The function does not undergo any abrupt changes at x_0 (it is continuous at the point x_0).
- It is independent of the direction in which the point x_0 is approached.

If the limit in equation (4.2) exists, it is called the **derivative** of the function $y = f(x)$ at the point x_0. The value of the derivative varies with the choice of x_0, and we define it in general terms as:

$$\left(\frac{dy}{dx} \right)_{x=x_0} = \lim_{\Delta x \to 0} \left\{ \frac{f(x_0 + \Delta x) - f(x_0)}{\Delta x} \right\} \qquad (4.3)$$

where $\left(\frac{dy}{dx} \right)_{x=x_0}$ is the name given to the value of the derivative at the point x_0. The derivative of the function $y = f(x)$ at $x = x_0$ in Figure 4.1 corresponds geometrically to the slope of the tangent to the curve $y = f(x)$ at the point P (known as the **gradient**).

The basic formula (4.3) for the derivative is often given in the form:

$$\frac{dy}{dx} = \lim_{\Delta x \to 0} \left\{ \frac{f(x + \Delta x) - f(x)}{\Delta x} \right\} \qquad (4.4)$$

for an arbitrary value of x.

We should also note that:

- $\frac{dy}{dx}$ is the name of the **derivative function**, commonly also represented as $f'(x)$.
- The domain of the derivative function is not necessarily the same as that of $y = f(x)$ (see Table 4.1).

The requirement that, for the limit in equation (4.2) to exist, the function does not undergo any abrupt changes is sometimes overlooked, yet it is an important one. An example of a function falling into this category is the modulus function, $y = |x|$, defined by:

$$y = f(x) = |x| = \begin{cases} x & \text{if } x \geqslant 0 \\ -x & \text{if } x < 0 \end{cases}$$

This function is continuous for all values of x (Figure 4.2a), but there is no unique slope at the point $x = 0$ as the derivative is undefined at this point (Figure 4.2b).

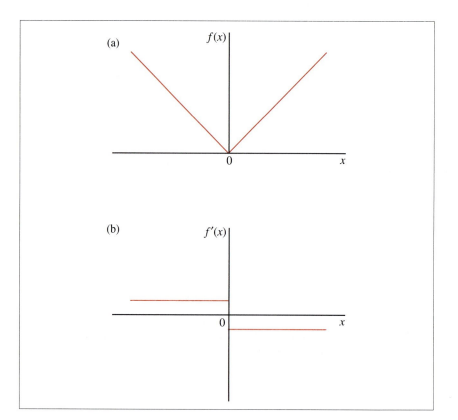

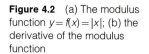
Figure 4.2 (a) The modulus function $y = f(x) = |x|$; (b) the derivative of the modulus function

Chemical examples showing this type of behaviour include processes associated with sudden changes in concentration, phase, crystal structure, temperature, *etc.* For example, Figure 2.9 shows how the equilibrium concentration of a chemical species changes suddenly when a temperature jump is applied at time t_0. Although there are no discontinuities in this function, its derivative is undefined at time t_0.

Worked Problem 4.1

Q Differentiate $y = f(x) = x^2$ using the definition of the derivative given in equation (4.4).

A $\dfrac{dy}{dx} = \lim\limits_{\Delta x \to 0} \left\{ \dfrac{f(x + \Delta x) - f(x)}{\Delta x} \right\} = \lim\limits_{\Delta x \to 0} \left\{ \dfrac{(x + \Delta x)^2 - x^2}{\Delta x} \right\}$

$\qquad = \lim\limits_{\Delta x \to 0} \left\{ \dfrac{x^2 + 2x\Delta x + (\Delta x)^2 - x^2}{\Delta x} \right\} = \lim\limits_{\Delta x \to 0} \left\{ \dfrac{2x\Delta x + (\Delta x)^2}{\Delta x} \right\}$

Since Δx tends to zero, but never takes the value zero, cancellation of Δx from all terms in the numerator and denominator yields:

$$\frac{dy}{dx} = \lim\limits_{\Delta x \to 0} \{2x + \Delta x\} = 2x$$

Problem 4.1

Differentiate the function $y = f(x)$, where $f(x) = 3$, using the definition of the derivative given in equation (4.4). *Hint*: the function $y = f(x) = 3$ requires that $y = 3$ for all values of x; thus if $f(x) = 3$, then $f(x + \Delta x)$ must also equal 3.

Problem 4.2

Use equation (4.4) to find the derivative of the function $y = f(x)$, where (a) $f(x) = 3x^2$ and (b) $f(x) = 1/x^2$. *Hint*: in your answer to (b), you will need to remember how to subtract fractions, *i.e.* $\frac{1}{a} - \frac{1}{b} = \frac{b-a}{ab}$.

4.2.2 Differentiation by Rule

Some Standard Derivatives

The derivatives of all functions can be found using the limit method described in Section 4.2.1. Some of the more common functions, and their derivatives, are listed in Table 4.1. Unless otherwise indicated, the respective domains (Dom) are "all values of x":

Table 4.1 Derivatives of some common functions, and their respective domains

$f(x)$	$f'(x)$	$\mathrm{Dom}(f(x))$	$\mathrm{Dom}(f'(x))$	Notes
c	0	–	–	1
x^n	nx^{n-1} $(n \neq 0)$	$x \neq 0$ for $n < 0$	$x \neq 0$ for $n < -1$	2
$\sin ax$	$a \cos ax$	–	–	3
$\cos ax$	$-a \sin ax$	–	–	4
$\tan ax$	$a \sec^2 ax$	$x \neq (2n+1)\pi/2$	$x \neq (2n+1)\pi/2$	5
$\sec ax$	$a \sec ax \tan ax$	$x \neq (2n+1)\pi/2$	$x \neq (2n+1)\pi/2$	6
$\ln ax$	a/x	$x > 0$	$x \neq 0$	7
e^{ax}	$a e^{ax}$	–	–	

[1] The constant function, c
[2] $n = 0$ corresponds to the constant function
[3] $a \neq 0$; for a 1:1 function, $\mathrm{Dom}(f(x)) = [-\pi/2, \pi/2]$
[4] $a \neq 0$; for a 1:1 function, $\mathrm{Dom}(f(x)) = [0, \pi]$
[5,6,7] $a \neq 0$

However, as we have seen above, and in Table 4.1, we do meet functions for which the derivative $f'(x)$ does not exist at selected values of x. The functions $y = f(x) = \ln x$ at $x = 0$ and $y = f(x) = \tan x$ at $x = (2n+1)\pi/2$, both listed in Table 4.1, fall into this category. Naturally, since the derivative does not exist in these cases at selective values of x, the domain of the derivatives of these functions will not be the same as the original functions. The restrictions on the respective domains are best seen in sample plots of these functions shown in Figure 4.3.

An Introduction to the Concept of the Operator

The notation $\frac{dy}{dx}$ (or sometimes dy/dx) for the derivative is just one of a number of different notations in widespread use, all of which are equivalent:

$$\frac{dy}{dx}, \quad dy/dx, \quad f'(x), \quad f^{(1)}(x), \quad \hat{D}f(x)$$

The more commonly used notations are $\frac{dy}{dx}$ and $f'(x)$, but expressing the derivative in the form $\hat{D}f(x)$ provides a useful reminder that the derivative function is obtained from the function $y = f(x)$ by the *operation* "differentiate with respect to x". Thus, we express this instruction in symbols as:

$$\hat{D}f(x) \equiv \frac{d}{dx}f(x) = \frac{d}{dx}y = \frac{dy}{dx} \tag{4.5}$$

It is worth emphasizing that the symbol $\frac{dy}{dx}$ does not mean dy divided by dx in this context, but represents the *limiting* value of the quotient $\Delta y/\Delta x$ as $\Delta x \to 0$.

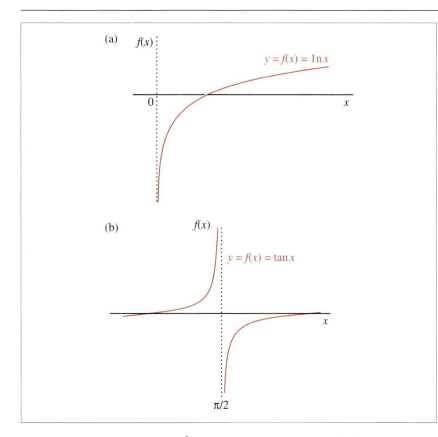

(a) $y = f(x) = \ln x$

(b) $y = f(x) = \tan x$

Figure 4.3 The functions (a) $y = f(x) = \ln x$ and (b) $y = f(x) = \tan x$ are both examples of functions for which the derivative does not exist at certain values in the independent variable (see Table 4.1)

In general, an **operator**, $\hat{A}$, is represented by a symbol with a caret ("hat") denoting an instruction to undertake an appropriate action on the object to its right (here $f(x)$). In equation (4.5), we consider $\frac{dy}{dx}$ to be the differentiation operator $\frac{d}{dx}$ acting on the function $f(x)$, which we have labelled y, to give a new function, say $g(x)$:

$$\hat{A}f(x) = g(x) \qquad (4.6)$$

Worked Problem 4.2

Q For the function $f(x) = x^2$, find $\hat{A}(f(x))$ where $\hat{A} = d/dx$.

A For $f(x) = x^2$, $\frac{d}{dx}(f(x)) = 2x$.

Problem 4.3

For each of the following functions, $f(x)$, use the information in Table 4.1 to find $\hat{A}(f(x))$, where $\hat{A} = d/dx$: (a) $x^{3/4}$; (b) e^{-3x}; (c) $1/x$; (d) $a \cos ax$.

> **Problem 4.4**
>
> Use the information in Table 4.1 to demonstrate that, when the operator $\hat{A} = \frac{d}{dx} + 2$ acts on $f(x) = e^{-2x}$, the function is annihilated (*i.e.* the **null function**, $g(x) = 0$, results).

We will come to appreciate the full significance of the concept of the operator in Section 4.3.1, when we consider the *eigenvalue problem*.

4.2.3 Basic Rules for Differentiation

Although all functions can be differentiated from first principles, using equation (4.4), this can be a rather long-winded process in practice. In this chapter, we deal with the differentiation of more complicated functions with the aid of a set of rules, all of which may be derived from the defining relation (4.4). In many cases, however, we simply need to learn what the derivative of a particular function is, or how to go about differentiating a certain class of function. For example, we learn that the derivative of $y = f(x) = \sin x$ is $\cos x$, but that the derivative of $y = f(x) = \cos x$ is $-\sin x$. Similarly, we can differentiate any function of the type $y = f(x) = x^n$ by remembering the rule that we reduce the index of x by 1, and multiply the result by n; that is:

$$\frac{d}{dx} x^n = nx^{n-1} \qquad (n \neq 0) \tag{4.7}$$

For functions involving a combination of other elementary functions, we follow another set of rules: if u and v represent functions $f(x)$ and $g(x)$, respectively, then the rules for differentiating a sum, product or quotient can be expressed as:

$$\frac{d}{dx}(u + v) = \frac{du}{dx} + \frac{dv}{dx} \tag{4.8}$$

$$\frac{d}{dx}(uv) = v\frac{du}{dx} + u\frac{dv}{dx} \tag{4.9}$$

$$\frac{d}{dx}\left(\frac{u}{v}\right) = \frac{v\frac{du}{dx} - u\frac{dv}{dx}}{v^2} \tag{4.10}$$

> **Problem 4.5**
>
> Differentiate the following, using the appropriate rules:
> (a) $(x-1)(x^2+4)$; (b) $\frac{x}{(x+1)}$; (c) $\sin^2 x$; (d) $x \ln x$; (e) $e^x \sin x$.

4.2.4 Chain Rule

Quite frequently we are faced with the problem of differentiating functions of functions, such as $y = \ln(x^2 + x + 1)$. The derivative of this function is not immediately obvious, and so we use a strategy known as the **chain rule** to reduce the problem to a more manageable form. We can proceed as follows:

- Introduce a new variable $u = x^2 + x + 1$ to transform the function $y = \ln(x^2 + x + 1)$ into the simpler form $y = \ln u$.
- Determine the derivative of y with respect to u:

$$\frac{dy}{du} = \frac{1}{u}$$

- Determine the derivative of u with respect to x:

$$\frac{du}{dx} = 2x + 1$$

- Combine the two derivatives, using:

$$\frac{dy}{dx} = \frac{dy}{du}\frac{du}{dx} = \frac{1}{u}(2x + 1)$$

- Eliminate the variable u:

$$\frac{dy}{dx} = \frac{2x + 1}{x^2 + x + 1}$$

Problem 4.6

Apply the chain rule to find the derivative of $y = e^{x\sin x}$, using the substitution $u = x \sin x$.

Problem 4.7

Use the chain rule to find the derivative of the following functions:
(a) $y = \ln(2 + x^2)$; (b) $y = 2\sin(x^2 - 1)$.

4.3 Higher-order Derivatives

In general, when we differentiate a function $y = f(x)$, another function of x is obtained:

$$\frac{dy}{dx} = f'(x)$$

If this derivative function is specified, say, by the relation $h = f'(x) = g(x)$, then, so long as $g(x)$ is not zero, h may be differentiated again to yield the *second derivative* of $f(x)$:

When we write $h = f'(x) = g(x)$, we are simply labelling the function $g(x)$ that results from differentiation of $f(x)$ arbitrarily with the letter h in the same way that we labelled $f(x)$ with y.

$$\frac{dh}{dx} = \frac{d}{dx} h = \frac{d}{dx}\frac{dy}{dx} = \frac{d^2 y}{dx^2} \equiv f''(x) \quad \text{or} \quad f^{(2)}(x) \qquad (4.11)$$

This process may usually be repeated to determine higher-order derivatives, if they exist. Thus, for example, if $f(x) = x^3 - x + 1$, then:

$$\frac{dy}{dx} = 3x^2 - 1, \ \frac{d^2 y}{dx^2} = 6x, \ \frac{d^3 y}{dx^3} = 6 \quad \text{and} \quad \frac{d^n y}{dx^n} = 0 \quad \text{for} \quad n > 3$$

Worked Problem 4.3

Q Given $y = f(x) = (1 + x)^4$, find $\dfrac{d^2 y}{dx^2}$. Deduce for what value of n, $\dfrac{d^n y}{dx^n} = 0$.

A Let $u = (1 + x)$ to transform the function into a simpler form, $y = u^4$, and use the chain rule to find $\frac{dy}{dx}$:

- $\dfrac{dy}{dx} = \dfrac{dy}{du}\dfrac{du}{dx} = 4u^3 \times 1 = 4u^3 = 4(1 + x)^3$

- Let $h = \dfrac{dy}{dx} = 4(1 + x)^3$ and use the chain rule again:

$$\frac{d^2 y}{dx^2} = \frac{dh}{dx} = \frac{dh}{du}\frac{du}{dx} = 12u^2 \times 1 = 12(1 + x)^2$$

In this example, we can see that each act of differentiation decreases the index of $(1 + x)$ by one and so it follows that the fifth derivative will be zero.

We can gain some useful insight into what exactly the first and second derivatives of a function tell us by looking at the form of the three functions $f(x)$, $f'(x)$ and $f''(x)$ shown in Figure 4.4.

The original function $y = f(x) = (1 + x)^4$ must be positive for all values of x and has a minimum value of zero at $x = -1$. The first derivative $f'(x) = 4(1 + x)^3$ gives us the rate of change (slope of the tangent) of the function $f(x)$ for any value of x. For $x < -1$ the value of $f'(x) = 4(1 + x)^3$ is negative, which means the slope of the original function is also negative (which we can see for ourselves by inspection of the plot). For $x > -1$ the first derivative is positive and so the slope of the original function is also

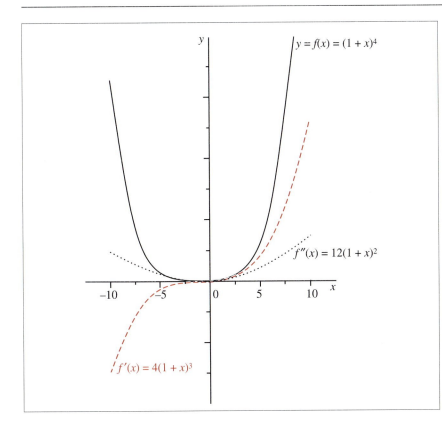

$y = f(x) = (1 + x)^4$

$f''(x) = 12(1 + x)^2$

$f'(x) = 4(1 + x)^3$

Figure 4.4 Plots of the function $f(x) = (1+x)^4$ and its first two derivatives

positive. The fact that the value of $f'(x) = 4(1 + x)^3$ is zero at $x = -1$ indicates that the slope of the function is zero at this point. Such a point is identified as a **stationary point**, which, in this case, corresponds to a minimum (as we can see from the plot). We shall see later in Section 4.4 how to prove whether a stationary point is a maximum or minimum (or point of inflection) without needing to plot the function. Similarly, the form of the second derivative, $f''(x) = 12(1 + x)^2$, gives us the slope, or rate of change, of the first derivative and by extension the slope of the slope of the original function $f(x)$. The form of the second derivative provides us with the means to characterize the nature of any stationary points in the original function, while that of the first derivative tells us if and where the stationary points exist (see Section 4.4).

Problem 4.8

Find the second and third derivatives of (a) $y = 1/x$ and (b) $y = N \sin ax$ (N, a are constants).

4.3.1 Operators Revisited: an Introduction to the Eigenvalue Problem

In Section 4.2.2 we defined the act of differentiation as an operation in which the operator $\hat{D} = \mathrm{d}/\mathrm{d}x$ acts on some function $f(x)$. Similarly, we can express the act of differentiating twice in terms of the operator $\hat{D}^2 = \mathrm{d}^2/\mathrm{d}x^2$.

Worked Problem 4.4

Q   For the function $f(x) = \cos kx$, find $\hat{A}(f(x))$, where $\hat{A} = \mathrm{d}^2/\mathrm{d}x^2$.

A   For $f(x) = \cos kx$, $\mathrm{d}/\mathrm{d}x(f(x)) = -k\sin kx$ and $\mathrm{d}^2/\mathrm{d}x^2(f(x)) = -k^2\cos kx$ and so $\hat{A}\cos kx = -k^2\cos kx$.

The Eigenvalue Problem

A problem common to many areas in physical chemistry is the following: given an operator, $\hat{A}$, find a function $\phi(x)$, and a constant a, such that $\hat{A}$ acting on $\phi(x)$ yields a constant multiplied by $\phi(x)$. In other words, the result of operating on the function $\phi(x)$ by $\hat{A}$ is simply to return $\phi(x)$, multiplied by a constant factor, a. This type of problem is known as an **eigenvalue problem**, and the key features may be described schematically as follows:

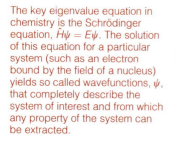

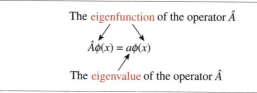

The eigenfunction of the operator $\hat{A}$

$$\hat{A}\phi(x) = a\phi(x)$$

The eigenvalue of the operator $\hat{A}$

The key eigenvalue equation in chemistry is the Schrödinger equation, $\hat{H}\psi = E\psi$. The solution of this equation for a particular system (such as an electron bound by the field of a nucleus) yields so called wavefunctions, ψ, that completely describe the system of interest and from which any property of the system can be extracted.

The solution to Worked Problem 4.4 is an example of an eigenvalue problem.

Worked Problem 4.4 *revisited*

For $f(x) = \cos kx$ and $\hat{A} = \mathrm{d}^2/\mathrm{d}x^2$:

$$\hat{A}\cos kx = -k^2\cos kx$$

In this example, we see that by differentiating the function $f(x) = \cos kx$ twice, we regenerate our original function multiplied by a constant which, in this case, is $-k^2$. Hence, $\cos kx$ is an eigenfunction of $\hat{A}$, and its eigenvalue is $-k^2$.

Problem 4.9

Perform the following operations:

(a) For $f(x) = x^3$, find, $\hat{A}(f(x))$, where $\hat{A} = d^2/dx^2$.

(b) For $f(x) = \sin kx$, find $\hat{A}(f(x))$, where $\hat{A} = d^2/dx^2$.

(c) For $f(x) = \sin kx + \cos kx$, find $\hat{A}(f(x))$, where $\hat{A} = d^2/dx^2$.

(d) For $f(x) = e^{ax}$, find $\hat{A}(f(x))$, where $\hat{A} = d/dx$.

Which of (a)–(d) would be classified as eigenvalue problems? What is the eigenfunction and what is the eigenvalue in each case?

Problem 4.10

Show that $y = f(x) = e^{mx}$ is an eigenfunction of the operator $\hat{A} = \dfrac{d^2}{dx^2} - 2\dfrac{d}{dx} - 3$, and give its eigenvalue. For what values of m does $\hat{A}$ annihilate $f(x)$?

Annihilation of a function implies that the null function is produced after application of an operator.

Problem 4.11

The lowest energy solution of the **Schrödinger equation** for a particle (mass m) moving in a constant potential energy (V), and in a one-dimensional box of length L, takes the form:

$$\psi = \sqrt{\frac{2}{L}} \sin \frac{\pi x}{L}$$

If we take V as the zero of energy, then ψ satisfies the Schrödinger equation:

$$-\frac{h^2}{8\pi^2 m}\frac{d^2\psi}{dx^2} = E\psi$$

Find an expression for the total energy E in terms of L and the constants π, m and h. *Hint*: you may have noticed that the expression above is an example of an eigenvalue problem where the eigenfunction is $\psi = \sqrt{\frac{2}{L}} \sin \frac{\pi x}{L}$ and the eigenvalue is E. In this case, the total energy E is determined by operating on the function ψ using the operator $-\frac{h^2}{8\pi^2 m}\frac{d^2}{dx^2}$

In quantum mechanics, the operator $-\frac{h^2}{8\pi^2 m}\frac{d^2}{dx^2}$ is called the **Hamiltonian** and is given the symbol $\hat{H}$.

4.4   Maxima, Minima and Points of Inflection

We often encounter situations in the physical sciences where we need to establish at which value(s) of an independent variable a maximum or minimum value in the function occurs. For example:

- The probability of finding the electron in the ground state of the hydrogen atom between radii r and $r + dr$ is given by $D(r)dr$, where $D(r)$ is the radial probability density function shown in Figure 4.5. The most probable distance of the electron from the nucleus is found by locating the maximum in $D(r)$ (see Problem 4.12 below). It should come as no surprise to discover that this maximum occurs at the value $r = a_0$, the Bohr radius.

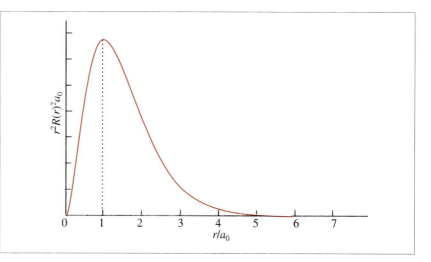

Figure 4.5 The radial probability density function for the 1s atomic orbital of the hydrogen atom

- When we attempt to fit a theoretical curve to a set of experimental data points, we typically apply a least-squares fitting technique which seeks to minimize the deviation of the fit from the experimental data. In this case, differential calculus is used to find the minimum in the function that describes the deviation between fit and experiment.

4.4.1 Finding and Defining Stationary Points

Consider the function $y = f(x)$ in Figure 4.6. As we saw in our discussion of Worked Problem 4.3, values of x for which $f'(x) = 0$ are called stationary points. A stationary point may be:

- A maximum (point E, a turning point) or a minimum (point C, a turning point). The value of dy/dx changes sign on passing through these points.
- A point of inflection: the tangent cuts the curve at this point (points A, B and D).

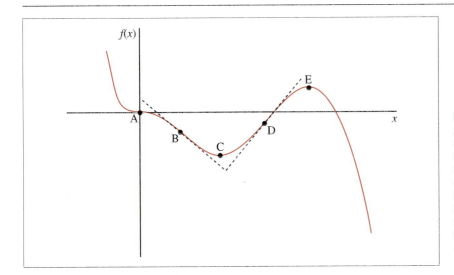

Figure 4.6 A plot of the function $y = f(x)$. Points A, C and E are all stationary points, for which $f'(x) = 0$, while points C and E are also turning points (minimum and maximum, respectively). Points A, B and D are all points of inflection, but B and D are neither stationary points nor turning points. Note that at the points of inflection, the tangents (*dashed lines*) cut the curve

Turning Points (Maxima and Minima)

Points E and C are called **turning points** because in passing through E and C the value of dy/dx changes sign. The existence and nature of stationary points, which are also turning points, may be identified through the first and second derivatives of the function. If we consider point C, we see that as we pass through this point the gradient becomes less negative as we approach C, passes through zero at point C, and then becomes positive. Clearly the *rate of change* of the gradient is positive at point C (because the gradient changes from negative to positive), which suggests that the function has a minimum at this point:

A **minimum** exists if $f'(x) = 0$ and $f^{(2)}(x) > 0$.

Similarly, on passing through point E, the gradient becomes less positive, passes through zero at E and then becomes negative. In this case, the rate of change in the gradient is negative and we can identify point E as a maximum:

A **minimum** exists if $f'(x) = 0$ and $f^{(2)}(x) < 0$.

In general, $y = f(x)$ will display a number of turning points within the domain of the function.

Turning points corresponding to maxima and minima may be classified as either:

- A **global maximum** or minimum which has a value greater or smaller than all other points within the domain of the function.
- A **local maximum** or minimum which has a value greater or smaller than all neighbouring points.

Points of Inflection

At a **point of inflection** (A, B, D), which may or may not be a stationary point:

- The tangent cuts the curve.
- The slope of the tangent does not change sign.

Note that A is both a point of inflection and a stationary point, but while B and D are both points of inflection, they are not stationary points because $f'(x) \neq 0$.

Points of inflection occur when the gradient is a maximum or minimum. This requires that $f^{(2)}(x) = 0$, but this in itself is not sufficient to characterize a point of inflection. We achieve this through the first non-zero higher derivative.

If $f'(x) = 0$, $f^{(2)}(x) = 0$ but $f^{(3)}(x) \neq 0$, then we have a point of inflection which is also a stationary point (such as point A). However, if $f'(x) \neq 0$, $f^{(2)}(x) = 0$ and $f^{(3)}(x) \neq 0$, then we have a point of inflection which is *not* a stationary point (B, D). The rules for identifying the location and nature of stationary points, turning points and points of inflection are summarized in Table 4.2.

Table 4.2 The location and nature of turning points, stationary points and points of inflection are given by the first, second and, where appropriate, third and fourth derivatives

	$f'(x)$	$f^{(2)}(x)$	$f^{(3)}(x)$	$f^{(4)}(x)$
Minimum	0	>0	–	–
Maximum	0	<0	–	–
Inflection point (stationary)	0	0	$\neq 0$	–
Inflection point (not stationary)	$\neq 0$	0	$\neq 0$	–
Turning points where $f^{(2)}(x) = 0$	0	0	0	$\neq 0$

Interestingly, in the last row of Table 4.2 we see that a turning point may exist for which $f^{(2)}(x) = 0$. In such cases, $f^{(3)}(x) = 0$, and the nature of the turning point is determined by the sign of the fourth derivative. An example of a function for which this latter condition applies is $y = f(x) = (x - 1)^4$. If there is any doubt over the nature of a stationary point, especially if the second derivative vanishes, it is always helpful to sketch the function!

Worked Problem 4.5

Q Consider the function $y = f(x)$, where $f(x) = x^2 - x^3/9$.

(a) Plot the function for selected values of x in the interval $-3.5 \leq x \leq 10$.

(b) Identify possible values of x corresponding to turning points and points of inflection.

(c) Derive expressions for the first and second derivatives of the function.

(d) Identify the nature of the turning points (*e.g.* maximum, minimum, global, local).

(e) Verify that there is a point of inflection where $f'(x) \neq 0$, $f^{(2)}(x) = 0$ and $f^{(3)}(x) \neq 0$.

A (a) See Figure 4.7.

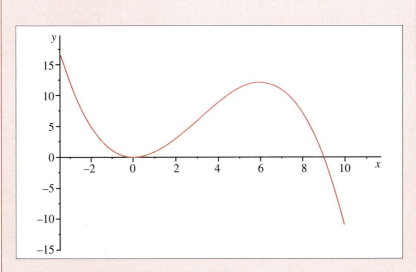

Figure 4.7 A plot of the function $y = f(x) = x^2 - x^3/9$ for $-3.5 \leq x \leq 10$

(b) By inspection, we can identify turning points at $x = 0$ and in the vicinity of $x = 6$; there is no turning point corresponding to a point of inflection.

(c) $f'(x) = 2x - x^2/3$; $f^{(2)}(x) = 2 - 2x/3$.

(d) $f'(x) = x\left(2 - \frac{x}{3}\right) = 0$ at $x = 0$ (local minimum; $f^{(2)}(x) > 0$) and $x = 6$ (local maximum; $f^{(2)}(x) < 0$).

(e) $f'(x) = 2x - x^2/3 \neq 0$ when $x = 3$; $f^{(2)}(x) = 2 - 2x/3 = 0$ when $x = 3$; $f^{(3)}(x) = -2/3 \neq 0$ when $x = 3$, corresponding to a point of inflection.

> ### Problem 4.12
>
> The radial probability density function for the electron in the ground state of the hydrogen atom takes the form $D(r) = Nr^2e^{-2r/a_0}$, where N is a constant.
>
> (a) Use the product rule to show that $dD(r)/dr = 2Ne^{-2r/a_0}(r - r^2/a_0)$.
>
> (b) Identify the non-zero value of r at which D displays a turning point, and give the value of D at this point.
>
> (c) Demonstrate, by examining the sign of the second derivative of D, that the turning point corresponds to a maximum.
>
> (d) Show that points of inflection, that are not stationary points, occur at $r = (1 \pm \frac{\sqrt{2}}{2})a_0$.

4.5   The Differentiation of Functions of Two or More Variables

In chemistry we frequently meet functions of two or more variables. For example:

- The pressure (p) of an ideal gas depends upon two independent variables, temperature (T) and volume (V):

$$p = \frac{nRT}{V}$$

- The electron probability density function, $\rho(x,y,z)$ for a molecule depends upon three spatial coordinates (x,y,z) to specify its value at a chosen position.
- The entropy, S, for a system containing three species, A, B and AB, at a given temperature and pressure depends upon five variables: N_A, N_B, N_{AB}, T and p, where N_X is the number of moles of A, B or AB.

When we explore the nature and form of these and other multi-variable functions, we need to know how to locate specific features, such as maximum or minimum values. Clearly, functions of two variables, such as in the ideal gas equation above, require plots in three dimensions to display all their features (such plots appear as surfaces). Derivatives of such functions with respect to one of these (independent) variables are easily found by treating all the other variables as constants and finding the **partial derivative** with respect to the single variable of interest.

For example, we can see from the ideal gas equation:

$$p = \frac{nRT}{V}$$

that p varies linearly with T but in a non-linear way with V (Figure 4.8).

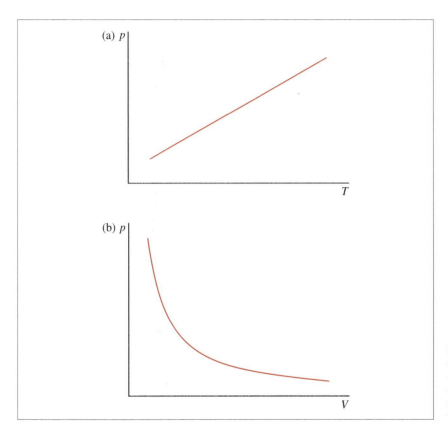

Figure 4.8 The ideal gas equation $p = nRT/V$ shows that (a) p varies linearly with T at constant volume but (b) in a non-linear way with V at constant temperature

If we were to differentiate this expression with respect to T and V, we would be able to evaluate precisely the rate at which p varied with respect to T (a constant) for any given T, or, with respect to V, for any given V (a variable) (see Problem 4.13).

Worked Problem 4.6

Q   For the function $z = xy + y^2$ find: (a) the partial derivative of z with respect to x; (b) the partial derivative of z with respect to y.

A   (a) The partial derivative of z with respect x is found by treating y as a constant. In order to make it clear that several variables are present, we sometimes use the notation $\left(\frac{\partial z}{\partial x}\right)_y$, which means differentiation with respect to x, keeping y constant. We often drop the y suffix and the brackets because there is usually no problem in recognizing which variables are held constant. In this case, the y^2 term is a constant and will vanish and so the derivative is given by $dz/dx = y$.

(b) Similarly, the partial derivative with respect to y, keeping x constant, is $dz/dy = x + 2y$.

Problem 4.13

For n mol of an ideal gas, $p = \frac{nRT}{V}$, so p is a function of the two variables T and V (R and n are constants). Write down $\partial p/\partial T$ and $\partial p/\partial V$.

Summary of Key Points

Differential calculus is inextricably linked to the notion of rates of change. This is especially important to our understanding of chemical kinetics and other areas of chemistry such as thermo-dynamics, quantum mechanics and spectroscopy. This chapter concerns the application of differential calculus to problems involving rates of change of one property with respect to another. The key points discussed include:

1. A comparison of average and instantaneous rates of change.

2. The use of limits to define the instantaneous rate of change as the derivative of a function.

3. Differentiation from first principles and by rule.

4. Differentiation of a sum, product or quotient of functions.

5. Discussion of operators, the eigenvalue problem and associated eigenfunctions.

6. The use of the chain rule for differentiating functions of functions.

7. Higher-order derivatives to locate and identify maxima, minima and points of inflection.

8. Differentiation of functions of more than one variable.

5
Differentials

In many areas of chemistry (*e.g.* error analysis; thermodynamics) we are concerned with the consequences of small (and, sometimes, not so small) changes in a number of variables and their overall effect upon a property depending on these variables. For example, in thermodynamics, the temperature dependence of the equilibrium constant, K, is usually expressed in the form:

$$K = e^{-\Delta G^{\circ}/RT}$$

where the change in Gibbs energy, $\Delta G^{\circ} = \Delta H^{\circ} - T\Delta S^{\circ}$, itself depends upon temperature, both explicitly through the presence of T, and implicitly, as ΔH° and ΔS° are, in general, both temperature dependent. However, if we assume that ΔH° and ΔS° are, to a good approximation, independent of temperature, then for small changes in temperature we obtain the explicit formula relating K and T:

$$K = e^{-(\Delta H^{\circ} - T\Delta S^{\circ})/RT} = e^{-(\Delta H^{\circ}/T - \Delta S^{\circ})/R} \tag{5.1}$$

Quite frequently, we are interested in the effect of *small changes* in the temperature on the equilibrium constant. We could, of course, use equation (5.1) to calculate K at two different temperatures for any reaction which satisfies the requirements given above and determine the change in K by subtraction. However, in practice, a much more convenient route makes use of the properties of differentials. This chapter is concerned with exploring what effect small changes in one or more independent variables have on the dependent variable in expressions such as equation (5.1). We shall see that this is particularly useful in determining how errors propagate through expressions relating one property to another. However, before discussing further the importance of differentials in a chemical context, we need to discuss some of the background to the method of differentials.

Aims

By the end of this chapter you should be able to:

* Understand the definition of change defined by the differential and the concept of infinitesimal change
* Understand the difference between the differential dy representing an approximate change in the dependent variable resulting from a small change in the independent variable, and the actual change in the dependent variable, Δy
* Calculate the differentials and the errors in approximating the differential to the actual change in a dependent variable
* Define the differential of a function of more than one variable
* Use differentials to calculate relative and percentage errors in one property deriving from those in other properties

5.1   The Effects of Incremental Change

We recall from Chapter 4 (Figure 4.1) that if Δy is the change in y that accompanies an *incremental* change Δx in x, then:

$$\Delta y = f(x + \Delta x) - f(x) \tag{5.2}$$

For example, if we consider the function $y = f(x) = x^3$, the incremental change in y that accompanies a change in Δx in x is given as:

$$\Delta y = (x + \Delta x)^3 - x^3$$

which, on expanding, yields:

$$\Delta y = 3x^2 \Delta x + 3x(\Delta x)^2 + (\Delta x)^3$$

For sufficiently small values of Δx, the power terms in Δx decrease very rapidly in magnitude. Thus, for example, if $\Delta x = 10^{-2}$, then $\Delta x^2 = 10^{-4}$ and $\Delta x^3 = 10^{-6}$. This may be expressed algebraically as:

$$(\Delta x)^3 << (\Delta x)^2 << \Delta x$$

and, if we neglect Δx raised to power 2 or higher, we can approximate the expression for Δy by:

$$\Delta y \approx 3x^2 \Delta x$$

The appearance of $3x^2$ in this expression is no accident. If we rewrite the expression for Δy as:

$$\Delta y = \left(\frac{f(x + \Delta x) - f(x)}{\Delta x} \right) \times \Delta x \tag{5.3}$$

then it is clear that, for very small Δx, the term in parentheses is an approximation for the derivative of $f(x)$, which, for the present choice of function, is $3x^2$. We can therefore rewrite the general result in the form $\Delta y \approx f'(x)\Delta x$.

5.1.1 The Concept of Infinitesimal Change

An **infinitesimal** change in x, known as the **differential** dx, gives rise to a corresponding change in y that is well represented by the differential dy:

$$dy = f'(x)dx \qquad (5.4)$$

The concept of an infinitesimal change is not soundly based mathematically: we interpret such changes as being very, very small (non-zero) increments in the specified variable.

We can see from the defining equation (5.4), and from Figure 5.1, that $f'(x)$ is the slope of the tangent to the curve $y = f(x)$ at the point P. We can also see that dy represents the change in the dependent variable y that results from a change, Δx, in x, as we move along the tangent to the curve at point P. It is important to stress that, although dy is *not* the same as Δy, for small enough changes in x it is reasonable to assume that the two are equivalent. Consequently, the difference between Δy and dy is simply the error in approximating Δy to dy. However, the same is not true of the differential dx, because, at all times, $\Delta x = dx$.

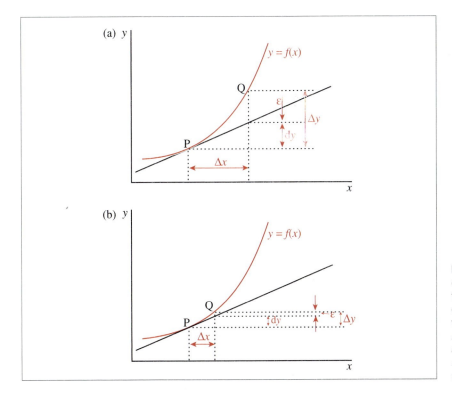

Figure 5.1 (a) The differential dy, for a change Δx in x, for the function $y = f(x)$. The actual change in y is given by $\Delta y = dy + \varepsilon$, where ε is the difference between Δy and dy. (b) As $\Delta x \to 0$, the error ε gets proportionately smaller and Δy becomes increasingly well approximated by dy.

The Origins of the Infinitesimal

The concept of the infinitesimal first arose in 1630 in Fermat's "Method of Finding Maxima & Minima". This work marks the beginning of differential calculus. The ideas introduced by Fermat lead to speculation about how we can evaluate "*just*" before or "*just*" after. In the 17th century, the infinitesimal was known as the "disappearing" and tangents as "touchings". Leibniz thought them "*useless fictions*", but they were subsequently recognized as being capable of producing extraordinary results. The philosopher Berkeley attacked differentials as "*neither finite quantities, nor quantities infinitely small, not yet nothing. May we not call them the ghosts of departed quantities*". Today, Borowski and Borwein in their Dictionary of Mathematics[1] regard an infinitesimal as "*a paradoxical conception · · · largely abandoned in favour of the epsilon-delta treatment of limits, · · · but made their reappearance in the formulation of hyper-real numbers*"!

5.1.2 Differentials in Action

The use of the differential is important in the physical sciences because fundamental theorems are sometimes expressed in differential form. In chemistry, for example, the laws of thermodynamics are nearly always expressed in terms of differentials. For example, it is common to work with the following formula as a means of expressing how the molar specific heat capacity at constant pressure, C_p, of a substance varies with temperature, T:

$$C_p = g(T) \text{ where } g(T) = \alpha + \beta T + \gamma T^2 \tag{5.5}$$

The optimum values of the parameters α, β, γ are found by fitting measured values of C_p over a range of temperatures to equation (5.5). Thus, if we know the value of C_p at one temperature, we can evaluate it at another temperature, and thereby determine the effect of that incremental (or decremental) change in temperature, ΔT, upon C_p, given by ΔC_p. Alternatively, we can use the properties of differentials given in equation (5.4) to evaluate the differential of C_p, dC_p, in terms of the differential dT as:

$$dC_p = g'(T)dT = (\beta + 2\gamma T) \times dT \tag{5.6}$$

The parameters in an expression such as equation (5.5) allow the expression to be tailored to fit experiment to some reasonable accuracy.

For small enough changes in T, it is reasonable to make the approximation that the differential dC_p is equivalent to the actual change ΔC_p, and we can use the expression above as a simple one-step route to evaluating the effect of small changes in T upon C_p.

Worked Problem 5.1

Q (a) Find dy and Δy for the function $y = f(x)$, where $f(x) = x^3$, given that $x = 4$ and $\Delta x = -0.1$. (b) Give the approximate and exact values of y at the point $x = 3.9$. (c) Calculate the percentage error in your approximate value from (b).

A (a) $f'(x) = 3x^2 \Rightarrow f'(4) = 48$. It follows that $dy = f'(4)\Delta x = 48 \times -0.1 = -4.8$. The actual change in y is given by $\Delta y = f(3.9) - f(4) = -4.681$.

(b) The actual and approximate values of y at $x = 3.9$ are 59.319 and 59.2, respectively.

(c) The percentage error is given by

$$\frac{59.319 - 59.2}{59.319} \times 100 = 0.201\%.$$

Sometimes, Δy will be smaller than dy, as in Worked Problem 5.1, but sometimes it can be larger: examples include functions whose slope decreases with increasing values of the independent variable, such as $y = f(x) = \ln x$ and $y = \sqrt[n]{x}$ where $n > 1$.

Problem 5.1

For the function $y = x^{1/3}$, find the values of the differential, dy, and the actual change, Δy, when the value of x is increased (a) from 27 to 30 and (b) from 27 to 27.1. Give the percentage error in each case in approximating Δy by dy.

Problem 5.2[2]

The variation of the molar heat capacity at constant pressure for $CH_4(g)$ is described by equation (5.5), with $\alpha = 14.143$ J K^{-1} mol^{-1}, $\beta = 75.495 \times 10^{-3}$ J K^{-2} mol^{-1} and $\gamma = -179.64 \times 10^{-7}$ J K^{-3} mol^{-1}.

(a) Use equation (5.5) to calculate the value of C_p at $T = 500$ K and at $T = 650$ K.

(b) Use equation (5.6) to evaluate dC_p for an incremental change in T, dT, of 150 K at $T = 500$ K. Hence, estimate the value of C_p at $T = 650$ K.

(c) Compare the value for C_p obtained in (b) with the value calculated directly from equation (5.5).

5.2 The Differential of a Function of Two or More Variables

We have seen in equation (5.4) that the differentials dy and dx are related through the derivative $dy = f'(x)dx$, which we can rewrite as:

$$dy = \frac{dy}{dx}dx \qquad (5.7)$$

We can now extend this principle to define differentials for functions of two or more variables. If $z = f(x, t)$ is a general function of two independent variables x and t, then there are two contributions to the differential dz: one from the change in x and the other from the change in t:

$$dz = \frac{\partial z}{\partial x}dx + \frac{\partial z}{\partial t}dt \qquad (5.8)$$

This result extends readily to functions of n independent variables x_1, x_2, x_3,..., x_n. Thus, if $z = f(x_1, x_2, x_3,..., x_n)$, the differential of z is built up from contributions associated with each independent variable, as a straightforward generalization of the result for two independent variables:

$$dz = \frac{\partial z}{\partial x_1}dx_1 + \frac{\partial z}{\partial x_2}dx_2 + ... + \frac{\partial z}{\partial x_n}dx_n = \sum_{i=1}^{n} \frac{\partial z}{\partial x_i}dx_i \qquad (5.9)$$

Examples of functions of two or more variables expressed in differential form are common in thermodynamics. For example, the equation:

$$dG = dH - TdS$$

relates the consequence of very small changes in the enthalpy, H, and entropy, S, on the Gibbs energy, G (here, G is the dependent variable, and H and S are the independent variables). As we shall see below, the use of differentials helps us to study such effects, if the changes are small. However, for large changes in the defining variables we have to evaluate the overall change in the property with the aid of *integral calculus*, which we meet in Chapters 6 and 7.

Worked Problem 5.2

Q Given the function $z = x^2y + y^2x - 2x + 3$, express dz in terms of dx and dy.

A $dz = \dfrac{\partial z}{\partial x}dx + \dfrac{\partial z}{\partial y}dy = (2xy + y^2 - 2)dx + (x^2 + 2xy)dy.$

Problem 5.3

If $z = xy/2$, express dz in terms of the differentials of the three independent variables.

Problem 5.4

(a) For a non-reacting system, the internal energy, $U = f(V, T)$, is a function of both V and T. By analogy with equation (5.8), write down an expression for the differential dU in terms of the differentials dV and dT.

(b) In thermodynamics, the expression derived in part (a) is commonly written as:[3]

$$dU = \pi_T dV + C_V dT$$

where π_T and C_V are the internal pressure and specific heat capacity at constant volume. (i) Use your answer to part (a) to find expressions for the internal pressure, π_T, and C_V. (ii) Assuming that $\Delta U \approx dU$, calculate the change in U that results when a sample of ammonia is heated from 300 K to 302 K and compressed through 100 cm^3, given that $C_V = 27.32$ J K^{-1} and $\pi_T = 840$ J m^{-3} at 300 K. Comment on the relative magnitudes of the two contributions to dU.

5.3   The Propagation of Errors

In many chemical situations we deduce a value for a property of interest by placing experimentally measured values in the right-hand side of an appropriate formula. For example, if we use the ideal gas equation:

$$p = n\frac{RT}{V} \tag{5.10}$$

to calculate the pressure, p, from a knowledge of volume, temperature, amount of substance and the gas constant, R, we might wish to know how the errors in the measured property values (n, T, V) propagate through to errors in the calculation of the pressure, p. If, for simplicity, we assume that n and R are fixed (given) constants, how can we estimate the error, dp, in p that results from errors, dT and dV, in the measurement of T and V, respectively? The answer lies in using equation (5.8) to obtain dp in terms of dV and dT:

$$dP = \frac{\partial P}{\partial T}dT + \frac{\partial P}{\partial V}dV \tag{5.11}$$

If dV and dT are the estimated errors in the measured values of V and T, then we need to know the two partial derivatives, so that we can estimate the error dp in P. However, in this and other instances the differentials themselves do not provide realistic measure of the errors. For example, an absolute error of 10 cm in a measured length is insignificant if we are talking about the shortest distance from Berlin to Moscow, but highly significant if a furniture van driver has enough clearance to pass under a low bridge in a country lane. For this reason, the **relative error**, or the closely related **percentage error**, give much more useful measures of error than absolute errors. Thus, in the context of the ideal gas example, the two kinds of error are defined as follows:

- The relative error in p is given by $\dfrac{dp}{p}$.

- The percentage error in p is given by $\dfrac{dp}{p} \times 100$.

Worked Problem 5.3

Q For a right-angled triangle with adjacent sides a, b and hypotenuse c, we have the relation $c = (a^2 + b^2)^{1/2}$. Find the relative and percentage errors in c when $a = 3$ cm, $b = 4$ cm, d$a = 0.1$ cm and d$b = 0.1$ cm.

A Using the chain rule, with the substitution $u = a^2 + b^2$, we initially define the partial derivatives of u with respect to a and b, respectively:

$$\frac{\partial u}{\partial a} = 2a; \quad \frac{\partial u}{\partial b} = 2b$$

Differentiating c with respect to the *single* variable, u, gives:

$$\frac{dc}{du} = \frac{1}{2}u^{-1/2}$$

Finally, we use the chain rule to obtain the partial derivatives of c with respect to a and b:

$$\frac{\partial c}{\partial a} = \frac{\partial u}{\partial a} \times \frac{dc}{du} = 2a \times \frac{1}{2}u^{-1/2} = 2a \times \frac{1}{2}(a^2 + b^2)^{-1/2} = a(a^2 + b^2)^{-1/2}$$

$$\frac{\partial c}{\partial b} = \frac{\partial u}{\partial b} \times \frac{dc}{du} = 2b \times \frac{1}{2}u^{-1/2} = 2b \times \frac{1}{2}(a^2 + b^2)^{-1/2} = b(a^2 + b^2)^{-1/2}$$

The differential dc is then given by:

$$dc = \frac{\partial c}{\partial a}da + \frac{\partial c}{\partial b}db = a(a^2 + b^2)^{-1/2}da + b(a^2 + b^2)^{-1/2}db$$

and so:

$$dc = 3(9 + 16)^{-1/2} \times 0.1 + 4(9 + 16)^{-1/2} \times 0.1 = 0.06 + 0.08$$
$$= 0.14 \text{ cm}$$

Thus the relative error

$$\frac{dc}{c} = \frac{0.14}{5} = 0.028$$

and the percentage error

$$\frac{dc}{c} \times 100 = 2.8\%.$$

Problem 5.5

The volume, V, of an orthorhombic unit cell with edges of length a, b and c and all internal angles between vertices of $90°$ is given by $V = abc$.

(a) Find the approximate change in volume, dV, when a, b and c change by da, db and dc, respectively.

(b) Give an expression for the percentage error in V, in terms of the percentage errors in a, b and c.

Problem 5.6

Calcium carbonate crystallizes in several different forms. In aragonite[4] there are four formula units in an orthorhombic primitive unit cell with dimensions $a = 4.94 \times 10^{-10}$ m, $b = 7.94 \times 10^{-10}$ m and $c = 5.72 \times 10^{-10}$ m.

(a) Calculate the mass, M, of a unit cell in kg, using molar atomic masses as follows:

$Ca = 40.08$ g mol^{-1}; $C = 12.01$ g mol^{-1}; $O = 16.00$ g mol^{-1} ($N_A = 6.022 \times 10^{23}$ mol^{-1}).

(b) Calculate the volume, V, of the unit cell, using the values of a, b and c above, and hence determine the density, ρ, of aragonite, using the formula $\rho = M/V$.

(c) Since the values of the unit cell parameters have been given to two decimal places, the error in their values is $\pm 0.005 \times 10^{-10}$ m. Ignoring the effects of the analogous errors associated with the masses of the atoms, give the relative and percentage errors in the volume of the unit cell.

(d) Find the greatest and smallest estimated unit cell volumes, and give the corresponding greatest and smallest estimates of the density (again ignoring errors associated with the relative atomic masses). Using the value of the density calculated in part (b), find the percentage errors and compare your answers to part (c).

Summary of Key Points

Differentials provide a means to quantify the effect of small changes in one or more variables upon a property that depends on those variables. The key points discussed include:

1. An illustration of the use of differentials in the mathematical and chemical context: in particular, many of the fundamental laws of thermodynamics are expressed in terms of differentials.

2. A review of the concept of infinitesimal change, and its relevance in chemistry, in view of the links to the concept of reversability in thermodynamics.

3. The distinction between approximate and exact changes in the dependent variable, resulting from changes in one or more independent variables.

4. The use of differentials in assessing how errors in one or more properties of a system propagate through to errors in a property that is related to those properties.

5. How differentials associated with each variable in a function of two or more variables contribute to the differential associated with the dependent variable.

References

1. E. J. Borowski and J. M. Borwein, *Collins Dictionary of Mathematics*, Harper Collins, New York, 1989, p. 294.
2. The data for Problem 5.2 were taken from R. A. Alberty and R. J. Silbey, *Physical Chemistry*, Wiley, New York, 1992, p. 52.
3. See, for example, P. W. Atkins, *Physical Chemistry,*5th edn., Oxford University Press, Oxford, 1994, p. 98.
4. See H. D. Megaw, *Crystal Structures: A Working Approach*, Saunders, Philadelphia, 1973, p. 247.

6
Integration

In the earlier chapters on arithmetic, algebra and functions, we saw examples of actions for which there was another action available to reverse the first action: such a reversing action is called an **inverse**. Some examples of mathematical actions and their inverses are listed below:

Start	→ Action	→ Result	→ Inverse action	→ Result
2	Add 3	5	Subtract 3	2
x^2	Subtract $2x$	$x^2 - 2x$	Add $2x$	x^2
$(x - 1)$	Multiply by x^3	$(x - 1)x^3$	Divide by x^3	$(x - 1)$
x	Logarithm	$\ln x$	Exponential $\exp(\ln x)$	x
$x^3 - x^2 + 1$	Differentiate	$3x^2 - 2x$	Integrate	$x^3 - x^2 + C$

Division by x^3 requires that $x \neq 0$.

The final example listed above proposes that the inverse to the operation of differentiation is known as **integration**. The field of mathematics which deals with integration is known as integral calculus and, in common with differential calculus, plays a vital role in underpinning many key areas of chemistry.

A differentiation/integration cycle involving a chosen initial function will lead to the appearance of an unspecified constant, C (as we shall see later on).

Aims

In this chapter we define and discuss integration from two perspectives: one in which integration acts as the inverse, or reverse, of differentiation and the other in which integration provides a means to finding the area under a curve. By the end of the chapter you should be able to:

- Understand the concept of integration as the reverse of differentiation
- Find the indefinite integral of a number of simple functions from first principles

- Integrate standard functions by rule
- Understand why the results of integration are not unique, unless constraints are placed on the integrated function
- Apply the integration by parts and substitution methods to integrate more complicated functions
- Understand the concept of the definite integral and be able evaluate a wide range of definite integrals using the methods discussed above

6.1   Reversing the Effects of Differentiation

Integration is used frequently in kinetics, thermodynamics, quantum mechanics and other areas of chemistry, where we build models based on changing quantities. Thus, if we know the rate of change of a property, y (the dependent variable), with respect to x (the independent variable), in the form of dy/dx, then integral calculus provides us with the tools for obtaining the form of y as a function of x. We see that integration reverses the effects of differentiation.

Consider, for example, a car undergoing a journey with an initial speed u and moving with a constant acceleration a. The speed, v, and distance, s, travelled after time t are given by:

$$v = u + at \text{ and } s = ut + \frac{1}{2}at^2 \tag{6.1}$$

The rate of change of distance with time yields the speed, v at time, t:

$$\frac{ds}{dt} = u + at = v \tag{6.2}$$

However, the reverse process, in going from speed to distance, involves integration of the **rate equation** (6.2). In chemistry, the concept of rate is central to an understanding of chemical kinetics, in which we have to deal with analogous rate equations which typically involve the rate of change of concentration, rather than the rate of change of distance. For example, in a first-order chemical reaction, where the rate of loss of the reactant is proportional to the concentration of the reactant, the rate equation takes the form:

$$-\frac{d[A]}{dt} = k[A] \tag{6.3}$$

where k, the constant of proportionality, is defined as the rate constant. The concentration of the reactant at a given time is found by integrating the rate equation (6.3), and the relationship between the differentiated and integrated forms of the rate equation is given schematically by:

$$-\frac{d[A]}{dt} = k[A]$$

differentiate ↑   ↓ *integrate*

$$[A] = [A]_0 e^{-kt}$$

where $[A]_0$ is the initial concentration of reactant A. We will discuss the integration methods required for obtaining the solution of this type of problem in some detail when we discuss differential equations in Chapter 7.

6.2 The Definite Integral

6.2.1 Finding the Area Under a Curve: The Origin of Integral Calculus

The concept of integration emerges when we attempt to determine the area bounded by a plot of a function $f(x)$ (where $f(x) > 0$) and the x axis, within an **interval** $x = a$ to $x = b$ (written alternatively as $[a,b]$). Clearly, if the plot gives a straight line, such as for the functions $y = 4$ or $y = 2x + 3$, shown in Figure 6.1, then measuring the area is straightforward, as the two areas are rectangular and trapezoidal in shape, respectively. However, for areas bounded by a curve and three straight lines, the problem is more difficult. The three situations are shown in Figure 6.1.

> The area of a trapezium is given by half the sum of the parallel sides, multiplied by the distance between them.

The solution to the general problem of determining the area under a curve arises directly from differential calculus, the concept of limits, and the infinitesimal. Seventeenth century mathematicians began to think of the area, not as a whole, but as made up of a series of rectangles, of width Δx, placed side by side, and which, together, cover the interval $[a,b]$ (see Figure 6.2).

With this construction, there are two ways of estimating the area under the curve. First, the interval $[a,b]$ is divided into n subintervals of width $\Delta x = (b - a)/n$. The area of each rectangle is obtained by multiplying its width, Δx, by its height on the left vertical side, as shown in Figure 6.3a.

In this case, the total area is given by:

$$A_1(n) = f(a)\Delta x + f(a + \Delta x)\Delta x + f(a + 2\Delta x)\Delta x + \ldots$$

$$\ldots + f(a + [n-1]\Delta x)\Delta x = \sum_{k=0}^{n-1} f(a + k\Delta x)\Delta x \quad (6.4)$$

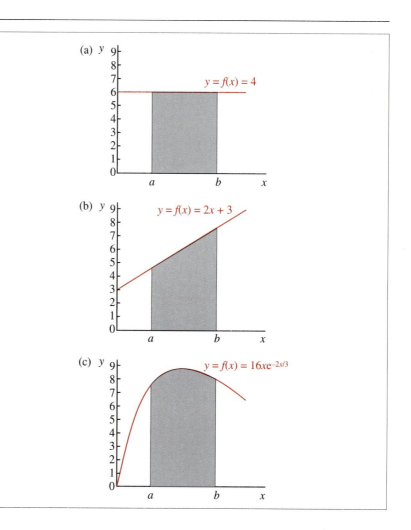

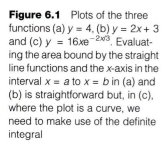

Figure 6.1 Plots of the three functions (a) $y = 4$, (b) $y = 2x + 3$ and (c) $y = 16xe^{-2x/3}$. Evaluating the area bound by the straight line functions and the x-axis in the interval $x = a$ to $x = b$ in (a) and (b) is straightforward but, in (c), where the plot is a curve, we need to make use of the definite integral

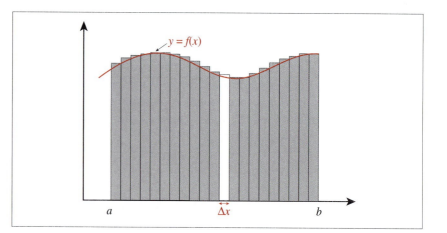

Figure 6.2 Approximating the area under a curve by a contiguous sequence of rectangles of width Δx

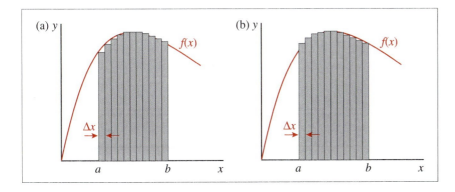

Figure 6.3 Choice of rectangles for estimating the area under the curve: (a) using the left side; (b) using the right side

Alternatively, we might have used the height of the right vertical side in computing the total area (Figure 6.3b), in which case the total area is given by:

$$A_r(n) = f(a + \Delta x)\Delta x + f(a + 2\Delta x)\Delta x + f(a + 3\Delta x)\Delta x + \ldots$$

$$\ldots + f(a + [n-1]\Delta x)\Delta x = \sum_{k=1}^{n} f(a + k\Delta x)\Delta x \qquad (6.5)$$

The two estimates we obtain for the area will be different, but if we decrease the subinterval width, thereby increasing the number, n, of subintervals, then in the limit $n \to \infty$, $A_l(n)$ and $A_r(n)$ converge to the same limiting value, A, which is the area under the curve:

$$A = \lim_{n \to \infty} A_l(n) = \lim_{n \to \infty} A_r(n) \qquad (6.6)$$

If $A_l(n)$ and $A_r(n)$ do not converge to the same value, A, then the integral is said to diverge, *i.e.* it is not defined.

Thus, from the definition of $A_r(n)$, and with an analogous expression involving the limit of $A_l(n)$, we can write:

$$A = \lim_{n \to \infty} \sum_{k=1}^{n} f(x_k)\Delta x \qquad (6.7)$$

where $f(x_k) = f(a + k\Delta x)$.

In order to symbolize this sum, Leibnitz introduced an elongated S which gives the familiar integral sign $\int$. Thus we can rewrite our equation as:

Note that the area, A, under the curve is the sum of an infinite number of rectangles of infinitesimally narrow width: an addition involving not a finite number of finite values, but an infinity of infinitesimal values, yielding a finite value!

$$A = \int_a^b f(x)\mathrm{d}x = \lim_{n \to \infty} \sum_{r=1}^{n} f(x_r)\Delta x_r \qquad (6.8)$$

where x takes all values between the lower and upper limits a and b, respectively. This integral is known as the **definite integral** because we have restricted x to the interval $[a,b]$ and, as seen in Figure 6.3, we can use the concept of area under the curve of $y=f(x)$ to give a visualization of the value of the integral.

Negative "Areas"

Attractive though the concept of area is when $f(x) \geq 0$, for x restricted to $[a,b]$, we do need to be careful if $f(x)$ also takes negative values in $[a,b]$. It turns out that, for those regions where the curve lies below the x axis, the contribution from $f(x)$ to the definite integral is negative. If it transpires that $A=0$, this is perfectly acceptable, as the definite integral has equal positive and negative contributions (see Figure 6.4); likewise, if the curve lies below the x axis, the definite integral will have a negative value.

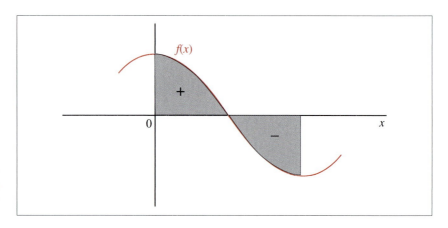

Figure 6.4   In this plot of the function $f(x) = \cos x$ the definite integral has a positive value over the interval $[0, \frac{\pi}{2}]$, a negative value over the interval $[\frac{\pi}{2}, \pi]$ and a zero value over $[0, \pi]$

6.2.2 A Chemical Example: Where is the Electron in the Hydrogen Atom?

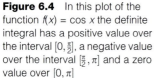

The radial probability density function is sometimes called the radial distribution function.

Consider the radial probability density function, $D(r)$, for the ground state of the hydrogen atom. This function describes the probability per unit length of finding an electron at a radial distance between r and $r + dr$ (see Figure 6.5).

The probability of finding the electron between r and dr is $D(r)dr$, and corresponds to the area under the curve between r and dr. Thus the area under the curve between $r=0$ and infinity simply gives us the probability of finding the electron somewhere in the interval 0 to ∞, which we know intuitively must be unity.

Before we discuss the definite integral any further, we first explore integration as the **inverse operation** to differentiation. This will prepare us for a most important result that enables us to evaluate the definite integral of $f(x)$, without first plotting the function as a prelude to computing the area under the curve.

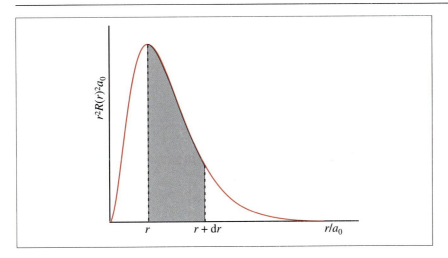

Figure 6.5 A plot of the radial probability density, $D(r) = Nr^2e^{-2r/a_0}$, for the 1s orbital of the hydrogen atom, where a_0 is the Bohr radius (units, m) and N has units of m^{-3}

6.3 The Indefinite Integral

The indefinite integral of a function $y = f(x)$ is usually written as:

$$\int f(x)\mathrm{d}x = F(x) + C \qquad (6.9)$$

where:

- $f(x)$ is known as the **integrand**.
- C is an arbitrary constant called the **constant of integration**.
- $F(x) + C$ is known as the **indefinite integral**.

The new function, $y = F(x) + C$, which we obtain after integration, must be such that its derivative is equal to $f(x)$, to ensure that the definition conforms with the requirement that integration is the reverse (or inverse) of differentiation. Thus, we must have:

$$\frac{\mathrm{d}}{\mathrm{d}x}(F(x) + C) = F'(x) = f(x) \qquad (6.10)$$

The relation between the indefinite integral of $f(x)$ and $f(x)$ itself is shown schematically in Figure 6.6 for the functions $f(x) = 18x^2$ and $F(x) = 6x^3$.

So, to summarize: the indefinite integral is determined by finding a suitable function, $F(x)$, which, on differentiation, yields the function we are trying to integrate, and to which we then add a constant. In common with the strategies described in Chapter 4 for finding the derivative of a given function, an analogous set of strategies can be constructed for finding the indefinite integral of a function. For simple functions, a set of standard indefinite integrals can be constructed without too much difficulty, some of which are listed in Table 6.1.

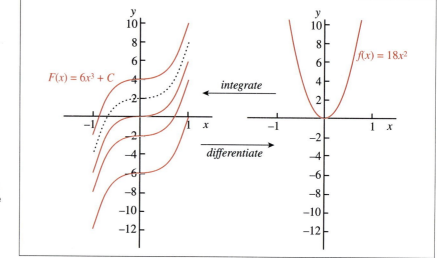

Figure 6.6 Integration of the function $f(x) = 18x^2$ (*right*) yields a family of functions given by the indefinite integral $F(x) = 6x^3 + C$ (*left*), where C can take any value. Differentiation of $F(x)$ yields the original function, $f(x)$

Table 6.1 A selection of functions, $f(x)$, and their indefinite integrals, $F(x) + C$

$f(x)$	integrate $\longrightarrow$	$F(x) + C$
$x^a\ (a \neq -1)$		$\frac{x^{a+1}}{a+1} + C$
$\frac{1}{x}$		$\ln(x), + C$
$\frac{1}{x+a}$		$\ln(x+a) + C$
$\cos(ax)$		$\frac{1}{a}\sin(ax) + C$
$\sin(ax)$		$-\frac{1}{a}\cos(ax) + C$
e^{ax}		$\frac{1}{a}e^{ax} + C$
$\sec^2(x)$		$\tan(x) + C$
$f(x)$	differentiate $\longleftarrow$	$F(x) + C$

Worked Problem 6.1

Q (a) Show that: $\frac{d}{dx}\ln(1 - 2x) = -\frac{2}{1-2x}$.

(b) Deduce that: $\int \frac{1}{1-2x}\,dx = -\frac{1}{2}\ln(1-2x) + C$.

A   (a) Since the first step involves establishing that the derivative of $y = \ln(1 - 2x)$ is $-\frac{2}{1-2x}$, it is simplest to use the chain rule (see Section 4.2.4). If $u = 1 - 2x$, then:

$$\frac{dy}{dx} = \frac{d}{dx}\ln(1-2x) = \frac{dy}{du}\frac{du}{dx} = \frac{1}{u}\times -2 = -\frac{2}{1-2x}$$

(b) Reversing the procedure by integration yields the following result, where B is the constant of integration:

$$-\int\frac{2}{1-2x}dx = \ln(1-2x) + B \Rightarrow \int\frac{1}{1-2x}dx = -\frac{1}{2}\ln(1-2x) + C$$

where $C = -B/2$.

Problem 6.1

(a) Evaluate $\frac{d}{dx}e^{2x}$ and hence deduce that $\int e^{2x}dx = \frac{1}{2}e^{2x} + C$.

(b) Show that $\frac{d}{dx}\left(\frac{1}{1+e^x}\right) = -\frac{e^x}{(1+e^x)^2}$ and hence find $\int\frac{e^x}{(1+e^x)^2}dx$.

6.4   General Strategies for Solving More Complicated Integrals

Integrals involving complicated forms for $f(x)$ require strategies for reducing the integral to one or more integrals of simpler (standard) form, thus making it possible to find $F(x)$. If all else fails, or we do not have an explicit form for $f(x)$, then numerical integration must be carried out, using methods described elsewhere.[1]

A chemical example of a function which does not have an explicit form can be found in thermodynamics, where the entropy is determined by integrating Cp/T, which may be known only at selected temperatures.

Some of the strategies involved in simplifying the form of an integral are quite straightforward. For example:

- If $f(x)$ is in the form of a linear combination of simpler functions, $e.g.$:

$$\int(3x^2 + 2x + 1)dx \tag{6.11}$$

then we may be able to rewrite such an integral as a sum of standard integrals that are immediately recognizable:

$$\int(3x^2 + 2x + 1)dx = \int 3x^2dx + \int 2xdx + \int 1dx \tag{6.12}$$

- Integrals can be simplified by placing constant terms outside the integral, $e.g.$:

$$\int(3x^2 + 2x + 1)dx = 3\int x^2dx + 2\int xdx + \int 1dx \tag{6.13}$$

Problem 6.2

Integrate the function $y = f(x) = 9x^2 + 2e^{2x} + \frac{1}{x}$.

In practice, we may find ourselves faced with more complicated functions, the solutions to which require us to use methods involving adaptation of some of the rules for differentiation. The choice of method more often than not involves some guesswork, but coming up with the correct guesses is all part of the fun! In addition, it may be necessary to use a combination of several methods. In the following two sections, we discuss **integration by parts** and the **substitution method**.

6.4.1 Integration by Parts

This method starts from the familiar product rule, used in differential calculus (equation 4.9):

$$\frac{d}{dx}(uv) = v\frac{du}{dx} + u\frac{dv}{dx}$$

Integration over x yields:

$$\int \frac{d}{dx}(uv)dx = \int v\frac{du}{dx}dx + \int u\frac{dv}{dx}dx \qquad (6.14)$$

and, on using the properties of differentials, the left side $\int \frac{d}{dx}(uv)dx$ becomes $\int d(uv) = uv$. It follows that rearrangement of the above expression yields:

$$\int u\frac{dv}{dx}dx = uv - \int v\frac{du}{dx}dx \qquad (6.15)$$

Equation (6.15) shows that the integral on the left side, which is the one sought, is replaced by two terms, one of which is another integral which we hope is more tractable than the initial integral. This method of integral evaluation is appropriate for integrands of product form. The success of the method relies on making the right choices for u and dv/dx. The term identified as u is differentiated to form part of the integrand on the right side of equation (6.15); the other part of the integrand is formed by integrating the term identified as dv/dx.

Worked Problem 6.2

Q   Given the integrand $f(x) = x \cos x$, find the indefinite integral.
A   The integrand is the product of x and $\cos x$, and in this case we identify x with u and dv/dx with $\cos x$ in equation (6.15): $u = x$ and

$dv/dx = \cos x$. Thus, $du/dx = 1$ and $v = \sin x$, and so equation (6.15) becomes:

$$\int x \cos x \, dx = x \sin x - \int \sin x \, dx.$$

The final step simply requires us to evaluate $\int \sin x \, dx$, which we know by reference to Table 6.1 to be $-\cos x + C$. Thus:

$$\int x \cos x \, dx = x \sin x + \cos x + D$$

where $D = -C$. If, on the other hand, we had identified u and dv/dx the other way round, we end up with a more complicated integral to evaluate:

$$\int x \cos x \, dx = x^2 \cos x + \int \frac{x^2}{2} \sin x \, dx$$

Clearly, some practice is required in identifying u and dv/dx for use in equation (6.15), when it seems that integration by parts is appropriate.

Problem 6.3

Use the method of integration by parts to evaluate $\int x e^{-x} dx$, assuming:

(a) $u = x$ and $dv/dx = e^{-x}$; (b) $u = e^{-x}$ and $dv/dx = x$.

Comment on which choice you think is the most appropriate for this integral.

6.4.2 Integration Using the Substitution Method

The second integration technique, known as the substitution method, derives from the inversion of the chain rule for differentiation described in Chapter 4. The objective here, once again, is to transform the integrand into a simpler or, preferably, a standard form. However, just like the integration by parts method, there is usually a choice of substitutions and although, in some cases, different substitutions yield different answers, these answers must only differ by a constant (remember that, for an indefinite integral, the answer is determined by inclusion of a constant). The substitution method is best illustrated using a worked problem.

Worked Problem 6.3

Q Evaluate $\int xe^{ax^2}dx$.

A Here $f(x) = xe^{ax^2}$. Let us try the substitution $u = ax^2$ in order to transform the integral over x to an integral over u. From the properties of differentials we know that:

$$du = \frac{du}{dx}dx = 2ax\,dx$$

This result enables us to express dx in terms of du, according to $dx = \frac{1}{2ax}du$, thus transforming the integral into:

$$\int xe^u \frac{du}{2ax} = \frac{1}{2a}\int e^u du = \frac{1}{2a}e^u + C$$

We now express the result in terms of the original variable, x, by substituting back for u:

$$\int xe^{ax^2}dx = \frac{1}{2a}e^{ax^2} + C$$

At this point, it is good practice to check the result by differentiating the function $F(x) = \frac{1}{2a}e^{ax^2}$, to ensure that the original integrand $f(x)$ is regenerated (see equation 6.10):

$$F'(x) = \frac{d}{dx}\left(\frac{1}{2a}e^{ax^2}\right) = \frac{2ax}{2a}e^{ax^2} = xe^{ax^2}$$

as required.

Problem 6.4

Repeat Worked Problem 6.3, using the substitution $u = x^2$.

Worked Problem 6.4

Q Evaluate

$$\int \frac{x}{(1-x)^{1/2}}dx.$$

A A possible substitution is given by $u = (1-x)^{1/2}$, from which it follows that:

$$u^2 = 1 - x \Rightarrow x = 1 - u^2$$

Differentiating the last equation with respect to u gives:

$$\frac{dx}{du} = -2u \Rightarrow dx = -2u\,du$$

Hence:

$$\int \frac{x}{(1-x)^{1/2}}\,dx = \int \frac{1-u^2}{u} \times -2u\,du$$

$$= -2 \int (1-u^2)\,du = -2u + \frac{2}{3}u^3 + C$$

$$\Rightarrow \int \frac{x}{(1-x)^{1/2}}\,dx = -\frac{2}{3}(1-x)^{1/2}(2+x) + C$$

Problem 6.5

Evaluate the indefinite integral $\int \dfrac{x}{(1-x)^{1/2}}\,dx$, using the substitution $u = 1-x^2$.

Problem 6.6

(a) Find $\int x(x^2+4)^{1/2}\,dx$, using the substitution $u = x^2 + 4$.

(b) Show that $\int \dfrac{1}{x\ln x}\,dx = \ln(\ln x) + C$, using the substitution $u = \ln x$.

Use of Trigonometrical Substitutions

The integrand in Problem 6.5 is of a form which suggests that a trigonometrical substitution might be appropriate. Bearing in mind the key identity $\cos^2 u + \sin^2 u = 1$, the appearance of a factor like $(1-x^2)^{1/2}$ in the integrand suggests the substitutions $x = \cos u$ or $x = \sin u$. Thus, for the substitution $x = \cos u$, the factor $(1-x^2)^{1/2}$ becomes $(1 - \cos^2 u)^{1/2} = \sin u$.

Problem 6.7

(a) Repeat Problem 6.5, using the trigonometrical substitution $x = \cos u$.

Hint. You will need to remember that $\sin^2 u = 1 - \cos^2 u$ and consequently that $\sin u = (1 - \cos^2 u)^{1/2}$ for the final step of your integration. You should have obtained the same result as your answer to Problem 6.5.

(b) Show that $\int \dfrac{\cos x}{\sin x} dx = \ln(\sin x) + C$ using the substitution $u = \sin x$.

General Comment

The choice of method for evaluating indefinite integrals relies on experience to a large extent. Sometimes, integration by parts and the substitution methods are equally applicable; however, in many cases they are not. For example, the integration by parts method is much more suited to finding the integral of the function $f(x) = x \cos x$, described in Worked Problem 6.2, than the substitution method (which would prove frustrating and fruitless in this case). It may also be necessary to use several applications of one or both methods before the answer is accessible. However, whichever method is used, the answer may always be checked by verifying that $F'(x) = f(x)$.

6.5 The Connection Between the Definite and Indefinite Integral

As we saw in Section 6.2.1, the concept of integration emerged from attempts to determine the area bounded by a plot of a function $f(x)$ and the x-axis, within some interval $[a,b]$. This area is given by the definite integral, the definition of which derives from numerical methods involving limits (see Section 6.2.1). Such numerical methods can be tedious to apply in practice (although instructive!) but, fortunately, there is a direct link between the indefinite integral, $F(x) + C$, of a function, $f(x)$, and the definite integral, in which x is constrained to the interval $[a,b]$. The relationship between the two forms of integration is provided by the fundamental theorem of calculus:

$$\int_a^b f(x)dx = (F(b) + C) - (F(a) + C) = F(b) - F(a) \qquad (6.16)$$

where $F(a)$ is the value of $F(x)$ at $x = a$ and $F(b)$ is the value of $F(x)$ at $x = b$. In other words, the definite integral over the interval $[a,b]$ is obtained by subtracting the indefinite integral at the point $x = a$ from that at $x = b$. Furthermore, we see that the constant of integration, which appears in the indefinite integral, does not appear in the final result (see equation 6.16).

Worked Problem 6.5

Q Evaluate $\int_0^1 \dfrac{x}{1+x}\,dx$.

A The first step requires us to find the indefinite integral $\int \frac{x}{1+x}\,dx$. Using the substitution $u = 1 + x$, the integral becomes:

$$\int \frac{u-1}{u}\,du = \int \left(1 - \frac{1}{u}\right)du = u - \ln u + C = (1+x) - \ln(1+x) + C$$

Thus identifying $F(x)$ with $(1 + x) - \ln(1 + x)$, the definite integral can be evaluated from:

$$\int_0^1 \frac{x}{1+x}\,dx = F(1) - F(0) = 2 - \ln 2 - 1 - 0 = 1 - \ln 2$$

Problem 6.8

(a) Evaluate (i) $\int_1^2 \dfrac{1}{x^3}\,dx$; (ii) $\int_0^2 x(x^2 + 4)^{1/2}\,dx$ (see Problem 6.6a).

(b) Show that $\int_0^2 \dfrac{x}{(x^2 + 4)}\,dx = \dfrac{1}{2}\ln 2$, using an appropriate substitution.

Problem 6.9

For the expansion of a perfect gas at constant temperature, the reversible work is given by the expression:

$$W = \int_{V_a}^{V_b} p\,dV$$

where $p = nRT/V$ and V_a and V_b are the initial and final volumes, respectively. Derive an expression for the work done by evaluating the integral between the limits V_a and V_b.

Problem 6.10

Let K be the equilibrium constant for the formation of CO_2 and H_2 from CO and H_2O at a given temperature T. From thermodynamics, we know that:

$$\frac{d}{dT} \ln K = \Delta H^{\ominus}/RT^2 \qquad (6.17)$$

(a) Assuming that $\Delta H^{\ominus}$ is independent of temperature, integrate equation (6.17) to find how $\ln K$ varies with T.

(b) Given $\Delta H^{\ominus} = 42.3$ kJ mol^{-1}, find the change in $\ln K$ as the temperature is raised from 500 K to 600 K.

Summary of Key Points

This chapter provides an introduction to integral calculus, together with examples set in a chemical context. However, as we shall see in the following chapter, we need integral calculus to solve the differential equations which appear in chemical kinetics, quantum mechanics, spectroscopy and other areas of chemistry. The key points discussed in this chapter include:

1. The definition of integration as the inverse of differentiation, yielding the indefinite integral.

2. The definition of integration as a means of evaluating the area bounded by a plot of a function over a given interval and the x-axis, yielding the definite integral.

3. The use of integration by parts method for integrating products of functions.

4. The use of the substitution method for reducing more complicated functions to a simpler or standard form.

5. The use of trigonometric substitutions in the substitution method.

Reference

1. See, for example, M. J. Englefield, *Mathematical Methods for Engineering and Science Students*, Arnold, London, 1987, chap. 15.

7
Differential Equations

In Chapter 2 we explored some of the methods used for finding the roots of algebraic equations in the form $y = f(x)$. In all of the examples given we were seeking to determine the value of an unknown (typically the value of the independent variable, x) that resulted in a particular value for y, the dependent variable. In general, the methods discussed can be used to solve algebraic equations where the dependent variable takes a value other than zero, because the equation can always be rearranged into a form in which $y = 0$. For example, if we seek the solution to the equation:

$$4 = x^2 - 5$$

then we can rearrange it to:

$$0 = x^2 - 9$$

by subtracting 4 from both sides. The problem now boils down to one in which we search for the two roots of the equation which, in this case, are $x = \pm 3$.

In this chapter we are concerned with equations containing derivatives of functions. Such equations are termed **differential equations**, and arise in the derivation of model equations describing processes involving rates of change, as in, for example:

- Chemical kinetics (concentrations changing with time).
- Quantum chemical descriptions of bonding (probability density changing with position).
- Vibrational spectroscopy (atomic positional coordinates changing with time).

In these three, as well as in other, examples we are trying to determine how the chosen property (such as concentration, probability density or atomic position) varies with respect to time, position or some other variable. This is a problem which requires the solution of one or more differential equations in a procedure that is made possible by using the tools of differentiation and integration discussed in Chapters 4 and 6, respectively.

Aims

This chapter builds on the content of earlier chapters to develop techniques for solving equations associated with processes involving rates of change. By the end of this chapter you should be able to:

- Identify a differential equation and classify it according to its order
- Use simple examples to demonstrate the origin and nature of differential equations
- Identify the key areas of chemistry where differential equations most often appear
- Use the separation of variables method to find the general solutions to first-order differential equations of the form $\frac{dy}{dx} = f(x)g(y)$
- Use the integrating factor method to find the general solutions to first-order differential equations linear in y
- Find the general solutions to linear second-order differential equations with constant coefficients by substitution of trial functions
- Apply constraints (boundary conditions) to the solution(s) of differential equations

7.1   Using the Derivative of a Function to Create a Differential Equation

Consider the function:

$$y = Be^{-2x} \tag{7.1}$$

where B is a constant. The first derivative of this function takes the form:

$$\frac{dy}{dx} = -2Be^{-2x} \tag{7.2}$$

If we now substitute for y, using equation (7.1), we obtain the **first-order** differential equation:

$$\frac{dy}{dx} = -2y \tag{7.3}$$

A first-order differential equation is so called because the highest order derivative is one.

which must be solved for y as a function of x. In other words, the solution to this problem will provide us with an equation which shows quantitatively how y varies as a function of x. The solution is, of course, provided by the original equation (7.1), but the purpose here is to explore the means by which we find that out for ourselves!

If we now differentiate equation (7.2) with respect to x, and substitute for dy/dx using equation (7.3), we obtain the second-order differential equation (7.4):

$$\frac{d^2y}{dx^2} = 4y \qquad (7.4)$$

This differential equation is of **second order**, simply because the *highest* order derivative is two.

Problem 7.1

(a) Express the first and second derivatives of the function $y = 1/x$ in the form of differential equations, and give their orders.

(b) Express the second derivative of the function $y = \cos ax$ in the form of a differential equation.

(c) Show that the function $y = Ae^{4x}$ is a solution of the differential equations $\frac{dy}{dx} - 4y = 0$ and $\frac{d^2y}{dx^2} - 5\frac{dy}{dx} + 4y = 0$

The last part of Problem 7.1 demonstrates that a given function does not necessarily correspond to the solution of only one differential equation. In later sections we shall address the question of how to determine the number of different functions (where each function differs from another by more than simply multiplication by a constant) that are solutions of a given differential equation.

If the function $y = Ae^{4x}$ is a solution to Problem 7.1(c), then so is ky, where k is a constant.

7.2 Some Examples of Differential Equations Arising in Classical and Chemical Contexts

One of the principal motivations for the development of calculus by Newton and Leibnitz in the 18th century came from the need to solve physical problems. Examples of such problems include:

- The description of a body falling under the influence of the force of gravity:

$$\frac{d^2h}{dt^2} = -g \qquad (7.5)$$

- The motion of a pendulum, which is an example of simple harmonic motion, described by the equation:

$$\frac{d^2x}{dt^2} = -\omega^2 x \qquad (7.6)$$

If a body is falling in a viscous medium, then the body is under the influence of both gravity and the drag forces exerted by the medium.

If we extend this last example to the modelling of molecular vibrations, we need to include additional terms in the differential equation to account for non-harmonic (anharmonic) forces.

In these last two examples of equations of motion, the objective is to determine functions of the form $h = f(t)$ or $x = g(t)$, respectively, which satisfy the appropriate differential equation. For example, the solution of the classical harmonic motion equation is an oscillatory function, $x = g(t)$, where $g(t) = \cos \omega t$, and ω defines the frequency of oscillation. This function is represented schematically in Figure 7.1 (see also Worked Problem 4.4).

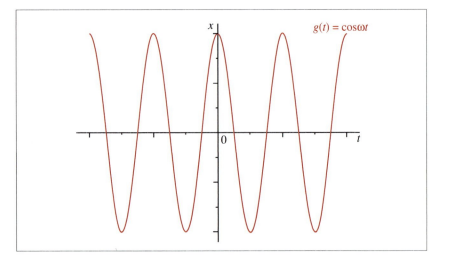

Figure 7.1 A plot of the function $g(t) = \cos \omega t$, describing simple harmonic motion

In chemistry, we are mostly concerned with changing quantities. For example:

- In kinetics, the concentration of a species A may change with time in a manner described by the solution of the differential equation:

$$-\frac{d[A]}{dt} = k[A] \tag{7.7}$$

- In quantum mechanics, the value of a wave function, ψ, changes with the position. For a single particle system, ψ is obtained as the solution of the Schrödinger equation:

$$-\frac{\hbar^2}{2m}\frac{d^2\psi}{dx^2} + V(x)\psi = E\psi \tag{7.8}$$

where the **Hamiltonian** operator, $\hat{H}$, given by $-\frac{\hbar^2}{2m}\frac{d^2}{dx^2} + V(x)$, is associated with the total energy, E, and $V(x)$ is the potential energy of the particle; m is the mass of the particle and $\hbar$ is the Planck constant divided by 2π.

- In spectroscopy, the response of a molecule to an oscillating electromagnetic field leads to absorption of energy, the details of which are revealed after solving an equation of the form:

$$i\hbar \frac{d\psi}{dt} = \left\{\hat{H} + \hat{H}'(t)\right\}\psi \qquad (7.9)$$

i is the imaginary number $\sqrt{-1}$ (see Chapter 2, Volume 2).

- In vibrational spectroscopy, where the treatment of molecular vibrations is based on the differential equation for an harmonic oscillator:

$$\frac{-\hbar^2}{2m}\frac{d^2\psi}{dx^2} + \tfrac{1}{2}kx^2\psi = E\psi \qquad (7.10)$$

In all of the examples given above, we are faced with having to deal with the relationship between some property and its rate of change. The differential equations that describe such relationships contain first-, second- or even higher-order derivatives. Most examples of this type of equation that we meet in chemistry are either of the first or second order, and so this is where we shall concentrate our efforts.

7.3   First-order Differential Equations

As already indicated, a *first-order* differential equation involves the *first* derivative of a function, and takes the general form:

$$\frac{dy}{dx} = F(x, y) \qquad (7.11)$$

where y is a function of x, and $F(x,y)$ is, in general, a function of *both* x and y. The method used to solve equation (7.11) depends upon the form of $F(x,y)$.

7.3.1 *F*(*x,y*) is Independent of *y*

In this simplest example, where $F(x,y) = f(x)$, the general solution is found by a simple one-step integration:

$$\frac{dy}{dx} = f(x) \Rightarrow y = \int f(x)dx = F(x) + C \qquad (7.12)$$

where $F(x) + C$ is the indefinite integral and C is the constant of integration (see Chapter 6), which can, in principle, take any value. It is important to note that the solution to a *first-order* equation involves:

- *One* step of integration.
- *One* constant of integration.

The solution of an nth order differential equation involves n steps of integration and yields n constants of integration.

Worked Problem 7.1

Q   Solve $\dfrac{dy}{dx} = x^2 + 1$.

A   Simple integration yields the general solution:

$$y = \frac{x^3}{3} + x + C$$

which can be described in terms of a family of cubic functions, each with a different value of C (see Figure 7.2). In this example, $F(x) = \dfrac{x^3}{3} + x$.

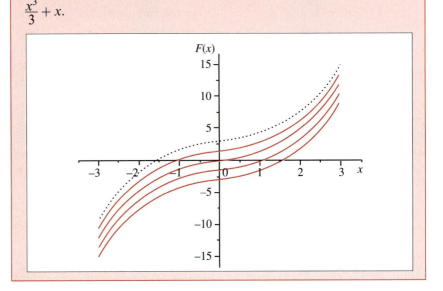

Figure 7.2   The family of solutions $y = \frac{x^3}{3} + x + C$ (for $C = \dots, 3, 1.5, 0, -1.5, -3, \dots$) to the differential equation $\frac{dy}{dx} = x^2 + 1$. Note that $F(x) = \frac{x^3}{3} + x$. The *dashed line* is the solution satisfying the boundary condition $y = 0$, $x = 3$

7.3.2 Boundary Conditions

In the case of a first-order differential equation, the constant of integration is usually determined by a **boundary condition**, or constraint on the solution. For example, if y is known at $x = 0$, then this boundary condition is sufficient to determine the constant of integration, C. Thus, out of the family of possible solutions, only one solution is acceptable and this is the one satisfying the boundary condition.

For example, if the boundary condition for the solution of the differential equation in Worked Problem 7.1 is such that $y = 3$ at $x = 0$, then the solution is constrained to take the form:

$$y = F(x) + 3 = \frac{x^3}{3} + x + 3$$

since $F(0) = 0 \Rightarrow C = 3$ (see the dashed-line solution in Figure 7.2).

It should be noted that, in most chemical situations, we rarely need the general solution of a differential equation associated with a particular property, because one (or more) boundary conditions will almost invariably be defined by the problem at hand and must be obeyed. For example:

- In a first-order reaction the concentration of the reacting species is specified at one particular time (usually at the start of the reaction).
- The value of the radial part of an atomic orbital wavefunction must tend to zero at very large distances from the nucleus.

As the number of boundary conditions is usually the same as the order of the differential equation for a particular chemical problem, there will be no undetermined constants of integration associated with the solution.

7.3.3 *F(x,y)* is in the Form *f(x)g(y)*

Separation of Variables Method

Suppose we are required to solve a differential equation of the form of equation (7.11), in which dy/dx is equal to the product of two functions, each of which depends only on one of the variables:

$$\frac{dy}{dx} = f(x)g(y) \tag{7.13}$$

This equation is solved by first rewriting it in the form:

$$\frac{1}{g(y)}\frac{dy}{dx} = f(x) \tag{7.14}$$

which, on integration with respect to x, yields:

$$\int \frac{1}{g(y)}\frac{dy}{dx}dx = \int f(x)\,dx \tag{7.15}$$

Since the differentials dx and dy are linked by the expression $dy = \frac{dy}{dx}dx$, the integration over x in the integral on the left-hand side of the equation can be transformed into an integration over y:

$$\int \frac{1}{g(y)}dy = \int f(x)\,dx \tag{7.16}$$

It now just remains to carry out separate integrations over y and x in order to obtain the required solution in the form of an expression of the general form:

$$G(y) + A = F(x) + B \tag{7.17}$$

Notice, that, although each integral yields *one constant of integration*, the two constants of integration can be combined into a *single constant*, C, after taking A over to the right side of equation (7.17).

The procedure just described is known as the **separation of variables method**. In some instances it is possible to re-write equation (7.17) in the form $y = P(x)$, to give an explicit relation between y and x [where B is contained within $P(x)$]; in other cases the solution may have to be left in a form of an implicit relation between y and x (see Section 2.3.5).

Worked Problem 7.2

Q   Find the solution of the differential equation: $dy/dx = 3x^2y$

A   This equation may be solved by the separation of variables method, as the right side is in the form $f(x)g(y)$, where $f(x) = 3x^2$ and $g(y) = y$. Thus:

$$\int \frac{1}{y} dy = \int 3x^2 dx \quad \Rightarrow \quad \ln y = x^3 + C$$

This expression gives y as an implicit function of x. However, from the properties of the exponential function (the inverse function of the logarithm function), we can specify y as an explicit function of x:

$$y = e^{(x^3 + C)} = e^C \times e^{x^3} \quad \Rightarrow \quad y = Ae^{x^3}$$

where the constant e^C has been rewritten as A (which implies $C = \ln A$), to simplify the appearance of the solution.

Problem 7.2

Solve the differential equation $dy/dx = -6y^2$, subject to the initial condition $y = 1$, $x = 0$.

Problem 7.3

Solve the differential equation $dy/dx = -\lambda y$, given that $y = N$ at $x = 0$. Give both implicit and explicit solutions.
Hint: you may find it helpful to remember that $\ln A - \ln B = \ln \frac{A}{B}$ (see equation 2.15) and that $e^{\ln A} = A$ (see equation 2.10).

7.3.4 Separable First-order Differential Equations in Chemical Kinetics

Consider a first-order rate process, with rate constant k:

$$A \xrightarrow{k} B \tag{7.18}$$

The rate of loss of the reactant A is proportional to its concentration, and is expressed in the form of the differential equation:

$$-\frac{d[A]}{dt} = k[A] \tag{7.19}$$

where $[A]$ is the concentration of the reactant at time t. Note here that $[A]$ is the dependent variable and t the independent variable.

We are interested in solving equation (7.19) to obtain an expression which describes how the concentration of A varies with time, subject to the boundary condition that the concentration of the reactant at time $t = 0$ is $[A]_0$ (note that the differential rate law above tells us only how the rate depends on $[A]$). Thus, using the separation of variables method, equation (7.19) is first rearranged to:

$$-\frac{d[A]}{[A]} = k\,dt \tag{7.20}$$

and then integrated, recognizing that k is a constant:

$$-\int \frac{d[A]}{[A]} = \int k\,dt = k\int dt \tag{7.21}$$

$$\Downarrow$$

$$-\ln[A] = kt + C \tag{7.22}$$

If we now impose the boundary condition above, we find that $C = -\ln[A]_0$, and the **integrated rate equation** (7.22) becomes:

$$-\ln[A] = kt - \ln[A]_0 \tag{7.23}$$

which may be expressed in the alternative forms:

$$\ln[A] = -kt + \ln[A]_0 \tag{7.23a}$$

or

$$\ln\left(\frac{[A]}{[A]_0}\right) = -kt \tag{7.23b}$$

Note that, in equation (7.23a), [A] is an implicit function of t; furthermore, there is a linear relation between $\ln[A]$ and t. Thus, a plot of $\ln[A]$ against t will give a straight line of slope $-k$ and intercept $\ln[A]_0$. Alternately, we can rearrange equation (7.23b) by taking the exponential of each side, to generate an explicit function which shows the exponential decay of [A] as a function of time (see Figure 7.3 and Chapter 2):

$$\frac{[A]}{[A]_0} = e^{-kt} \tag{7.24}$$

which rearranges to:

$$[A] = [A]_0 e^{-kt} \tag{7.25}$$

Figure 7.3 demonstrates clearly how the value of k determines the rate of loss of A.

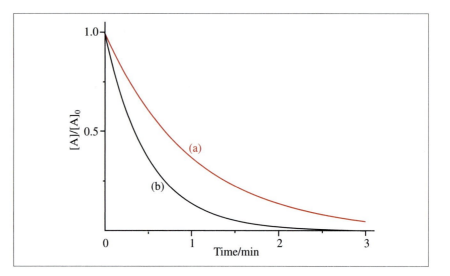

Figure 7.3 A plot of $\frac{[A]}{[A]_0}$ against t/min for (a) $k = 1$ min^{-1} and (b) $k = 2$ min^{-1}

An important feature of such first-order reactions is the half-life, $t_{1/2}$, which is the time taken for [A] to reduce to half of its initial value. Thus, for $t = t_{1/2}$, we have:

$$\frac{[A]_0}{2} = [A]_0 e^{-kt_{1/2}} \tag{7.26}$$

which simplifies to:

$$\frac{1}{2} = e^{-kt_{1/2}} \tag{7.27}$$

Taking natural logarithms, we have:

$$\ln \frac{1}{2} = -kt_{1/2} \tag{7.28}$$

Remember that $\ln(\frac{a}{b}) = \ln(a) - \ln(b)$.

Using the property of logarithms that $\ln \frac{1}{a} = -\ln a$, we can rewrite equation (7.28) as:

$$\ln 2 = kt_{1/2} \tag{7.29}$$

and it follows that the half-life is expressed in terms of the rate constant, k, according to:

$$t_{1/2} = \frac{\ln 2}{k} \tag{7.30}$$

7.3.5 First-order Differential Equations Linear in *y*

A first-order linear differential equation has the general form:

$$\frac{dy}{dx} + yP(x) = Q(x) \tag{7.31}$$

in which the dependent variable (here y) appears on the left-hand side with index unity. Equations of this form cannot be solved using the separation of variables method unless $Q(x) = 0$.

The general solution of a first-order linear differential equation, in the form of equation (7.31), is:

$$y = \frac{1}{R(x)} \int R(x)Q(x)\,dx \tag{7.32}$$

where $R(x)$, known as the **integrating factor,** is defined in terms of $P(x)$ as follows:

$$R(x) = e^{\int P(x)dx} \tag{7.33}$$

There are thus two integrations to perform: one to determine the integrating factor, and a second which involves the product $R(x)Q(x)$ as integrand. Since we are dealing with a first-order differential equation, we expect only *one* constant of integration, but, from the above discussion, it appears that *two* such constants may arise. We now describe why there is, in fact, only one undetermined constant of integration.

The Constant of Integration

In determining the integrating factor, the complete expression becomes:

$$R(x) = e^{g(x)+C} = Ae^{g(x)} \tag{7.34}$$

where $g(x) + C$ is the indefinite integral of the function $P(x)$ and $A = e^C$. Thus, if we now substitute equation (7.34) into (7.32), we obtain:

$$y = \frac{1}{Ae^{g(x)}} A \int e^{g(x)} Q(x) \, dx \qquad (7.35)$$

Since the constant A appears in both the numerator and denominator in the right side of equation (7.35), it can be cancelled to yield the general solution of equation (7.31):

$$y = e^{-g(x)} \int e^{g(x)} Q(x) \, dx \qquad (7.36)$$

Only one constant of integration will be produced from the indefinite integral $\int e^{g(x)} Q(x) \, dx$ and, since the constant arising from the determination of $R(x)$ can be discarded, we see that a single constant of integration arises from the solution to a first-order differential equation, as expected.

Worked Problem 7.3

Q   Solve the differential equation $\frac{dy}{dx} + 2y = e^x$.

A   This is a first-order linear differential equation, in which $P(x) = 2$ and $Q(x) = e^x$. The integrating factor $R(x) = e^{\int P(x) dx} = e^{\int 2 dx} = e^{2x+C}$. Here $g(x) = 2x$, and so the general solution given by equation (7.36) is:

$$y = e^{-2x} \int e^{2x} e^x dx = e^{-2x} \int e^{3x} dx$$

$$= e^{-2x} \left(\frac{e^{3x}}{3} + C \right) \Rightarrow y = \frac{1}{3} e^x + C e^{-2x}$$

Problem 7.4

(a) Check that the solution to Worked Problem 7.3 satisfies the original differential equation.
(b) Solve the differential equation $\frac{dy}{dx} + \frac{y}{x} = x^2$, subject to the boundary condition $y = 0$, $x = 1$.

7.3.6   First-order Differential Equations in Radioactive Decay Processes

Consider the following radioactive β^- decay processes, involving two sequential first-order steps, in which λ_1 and λ_2 are decay constants (analogous to rate constants in a chemical kinetic process) associated with the emission of energetic electrons:

$$^{239}_{92}\text{U} \xrightarrow{\lambda_1} {}^{239}_{93}\text{Np} \xrightarrow{\lambda_2} {}^{239}_{94}\text{Pu} \tag{7.37}$$

The amounts of $^{239}_{92}\text{U}$, $^{239}_{93}\text{Np}$ and $^{239}_{94}\text{Pu}$ (units mol) at any given time are denoted by N_1, N_2 and N_3, respectively, and we specify that, initially, $N_1 = 1$ mol. The change in the amount of $^{239}_{93}\text{Np}$ with time has two contributions: one from the decay of $^{239}_{92}\text{U}$ and the other from the decay of $^{239}_{93}\text{Np}$ itself. Thus, on the basis that these processes are first order in nature, the net rate of increase of $^{239}_{93}\text{Np}$ is given by:

$$\frac{dN_2}{dt} = \lambda_1 N_1 - \lambda_2 N_2 \tag{7.38}$$

By analogy with the first-order chemical reaction (equation 7.25) we know that:

$$N_1 = (N_1)_0 e^{-\lambda_1 t}$$

where $(N_1)_0$ is the initial amount of $^{239}_{92}\text{U}$. Thus:

$$\frac{dN_2}{dt} = \lambda_1 (N_1)_0 e^{-\lambda_1 t} - \lambda_2 N_2 \Rightarrow \frac{dN_2}{dt} + \lambda_2 N_2 = \lambda_1 (N_1)_0 e^{-\lambda_1 t} \tag{7.39}$$

If we identify t with x and N_2 with y, then we can see that equation (7.39) is of the form of equation (7.31), where $P(x) \equiv \lambda_2$ and $Q(x) \equiv \lambda_1 (N_1)_0 e^{-\lambda_1 t}$. After determining the integrating factor, the solution is obtained using equation (7.36). The derivation of the solution forms the basis of the next Problem.

^{239}U is produced when ^{238}U, the most common isotope of uranium, absorbs a neutron. The subsequent sequential decay process produces ^{239}Pu, a fissionable material that can be used as a fuel in nuclear reactors or as the core material of a nuclear bomb. 1 mol of $^{239}_{92}\text{U}$ equates to 239 g by mass, which is approximately 1/20 of the amount needed to make a nuclear device.

Problem 7.5

For the radioactive two-step decay process described above:
(a) Show that the integrating factor takes the form $R(t) = Ae^{g(t)}$, where $g(t) = \lambda_2 t$.
(b) Use equation (7.36) to show that the general solution is given by:

$$N_2 = \frac{\lambda_1 (N_1)_0}{\lambda_2 - \lambda_1} e^{-\lambda_1 t} + Ce^{-\lambda_2 t}$$

(c) Given the initial condition $N_2 = 0$ at $t = 0$, determine an expression for C, and show that:

$$N_2 = \frac{\lambda_1(N_1)_0}{\lambda_2 - \lambda_1}\left(e^{-\lambda_1 t} - e^{-\lambda_2 t}\right) \qquad (7.40)$$

(d) Given that the rate of loss of $^{239}_{92}U$, by first-order decay, is expressed in the form of the differential equation $\frac{dN_1}{dt} = -\lambda_1 N_1$, deduce the solution of this differential equation from the solution to Problem 7.3, by appropriate changes of names of the independent and dependent variables.

(e) Write down the expressions for N_1 and N_2 determined in parts (d) and (c) above, and hence, from the conservation of matter (*i.e.* that at all times $N_1 + N_2 + N_3 = 1$ mol), deduce the expression for N_3 in terms of time, t.

(f) Given that the half-lives for $^{239}_{92}U$ and $^{239}_{93}Np$ are 23.5 minutes and 2.3 days, respectively, use equation (7.30), with the rate constant k replaced by the appropriate decay constants, to calculate the values of the decay constants λ_1 and λ_2.

(g) Use the expression for N_2 derived in part (c) to determine the time at which N_2 reaches its maximum value, and give the corresponding amount of N_2 reached at this time.

In Problem 7.5 the large disparity in the decay constants leads to a situation in which the number of $^{239}_{93}Np$ species builds up rapidly to its maximum value, and then decreases slowly (see Figure 7.4).

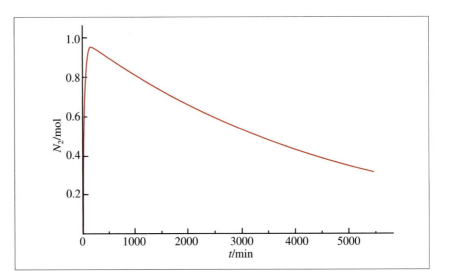

Figure 7.4 A plot of the variation in the amount of $^{239}_{93}Np$, given by N_2 as a function of time in the radioactive decay of $^{239}_{92}U$

7.3.7 First-order Differential Equations in Chemical Kinetics Processes

Consider the following kinetic process, involving two sequential first-order steps:

$$A \xrightarrow{k_1} B \xrightarrow{k_2} C$$

This is the same model process that we described above for radio-active decay of $^{239}_{92}U$; if we substitute decay constants by rate constants, and amount of substance by concentration, and assume that $[A]_0 = 1$ mol dm^{-3}, we can adapt equation (7.40) derived in Problem 7.5(c) to describe how [B] varies with time:

$$[B] = \frac{k_1}{k_2 - k_1}(e^{-k_1 t} - e^{-k_2 t})[A]_0 \qquad (7.41)$$

If we consider initially the limiting case, in which the rate constant k_2 (governing the second step) is very much smaller than k_1 (e.g. $k_1 = 2$ s^{-1}, $k_2 = 0.01$ s^{-1}), we obtain a plot of [B] over the time interval from 0 to 30 seconds shown in Figure 7.5.

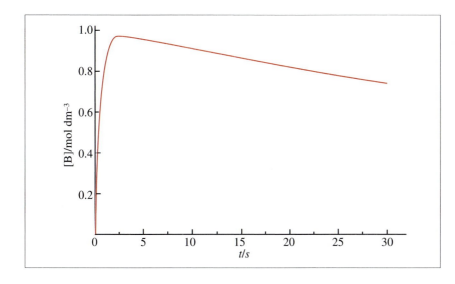

Figure 7.5 A plot of the variation in the concentration of the reaction intermediate, [B], with time in a two sequential first-order reaction mechanism

We can see in Figure 7.5 that, at the start of the reaction, the concentration of the intermediate, B, initially rises quite rapidly to a maximum, and thereafter declines slowly. The level to which the concentration of B builds up will depend on the rate constants governing the two steps. For reactions where the rate constant k_2 is very much larger than for k_1, the concentration of B never has the opportunity to build up and remains essentially constant throughout most of the reaction. Under such conditions it is said to have reached a steady state.

Problem 7.6

(a) Use equation (7.41) to plot [B]/mol dm^{-3} against t/s (see hint below), using first $k_1 = 2$ s^{-1}, $k_2 = 3$ s^{-1} and second $k_1 = 2$ s^{-1}, $k_2 = 10$ s^{-1}.

(b) Comment on the maximum values of [B] obtained in Figure 7.5, and in your plots generated in part (a).

Hint. If possible, you should use a computing plotting program to generate your plots. If you do not have access to such a program, a plotter is available on the Web at

http://www.ucl.ac.uk/mathematics/geomath/plot.html

In order to use such a plotter, you will need to enter the model expression for [B]/[A]$_0$ into a function descriptor box. Thus, in order to generate the plot shown in Figure 7.5, we would enter an expression such as:

$$(2/(0.01 - 2)) * (\exp(-2 * x) - \exp(-0.01 * x)),$$

where x is the appropriate name given to the independent variable in the plotter, and * is interpreted as multiplication.

7.4   Second-order Differential Equations

In Section 7.1 we saw how differential equations of second order can be generated from a particular function. Thus, for example, if we differentiate the function $y = e^{2x}$ twice, we obtain first:

$$\frac{dy}{dx} = 2y \tag{7.42}$$

and then:

$$\frac{d^2y}{dx^2} = 4y \tag{7.43}$$

which is a second-order differential equation. If we now reverse this process, we need two acts of integration to yield an expression for y. However, as each integration step leads to a constant of integration, the resulting expression for y, which now contains two undetermined constants, represents the **general solution** of the second-order differential equation (7.43).

In general, a second-order differential equation will take the form:

$$\frac{d^2y}{dx^2} = G\left(\frac{dy}{dx}, y, x\right) \tag{7.44}$$

where $G(\frac{dy}{dx}, y, x)$ can, in principle, represent any function of $\frac{dy}{dx}$, y and x. In the example above, equation (7.43) is not, in fact, a function of $\frac{dy}{dx}$ or x, and so we can write instead that equation (7.43) has the form $\frac{d^2y}{dx^2} = G(y)$.

Most of the problems involving second-order differential equations which we encounter in chemistry are **second-order linear differential equations**, which take the general form:

$$\frac{d^2y}{dx^2} + P(x)\frac{dy}{dx} + S(x)y = Q(x) \tag{7.45}$$

Unfortunately, unlike the general linear *first-order* differential equation (7.31), there is no simple template which provides the solution, and we need therefore to apply different methods to suit the equation we meet in the chemical context. Equations of the general form given in equation (7.45) crop up in all branches of the physical sciences where a system is under the influence of an oscillatory or periodic change. In chemistry, some of the most important examples can be found in modelling:

- Vibrational motions of molecules.
- The interaction of molecules with electromagnetic radiation (light).
- The radial motion of the electron in a hydrogen-like species.

As a first example, we will consider the differential equation describing the dynamics of simple harmonic motion, and demonstrate how the general solution is found.

7.4.1 Simple Harmonic Motion

The special case of equation (7.45) with $P(x) = Q(x) = 0$ and $S(x)$ equal to a positive constant, n^2 (the choice of n^2 as the constant ensures that it is positive quantity for any real value for n), gives rise to equation (7.46) for simple harmonic motion, the solution of which can be used to model nuclear motion in molecules:

$$\frac{d^2y}{dx^2} = -n^2y \tag{7.46}$$

Thus, for example, if we apply equation (7.46) to describe the periodic vibrational motion in a diatomic molecule, x represents time, and positive and negative values for y correspond to bond extension and compression, respectively. Finally, we can see that equation (7.46) is an eigenvalue equation in which y is the eigenfunction and $-n^2$ is the eigenvalue (see Section 4.3.1).

Worked Problem 7.4

Q   The form of equation (7.46) implies that y is such that its second derivative is a negative multiple of itself (since n^2 is positive real number).

(a) Which of the following functions are solutions of equation (7.46):
   (i) $y = \sin nx$; (ii) $y = \cos nx$; (iii) $y = e^{nx}$; (iv) $y = e^{-nx}$.

(b) Explain why none of these functions corresponds to the general solution of equation (7.46).

(c) Give the form of the general solution to equation (7.46).

(d) Find the solution to equation (7.46), given the boundary condition $y = 0$ when $x = 0$.

(e) If we additionally impose the boundary condition $y = 0$ when $x = L$, show that the solution becomes $y = A \sin \frac{n\pi x}{L}$, where A is a constant.

A   (a) (i) $\dfrac{dy}{dx} = n \cos nx \Rightarrow \dfrac{d^2 y}{dx^2} = -n^2 \sin nx = -n^2 y.$

(ii) $\dfrac{dy}{dx} = -n \sin nx \Rightarrow \dfrac{d^2 y}{dx^2} = -n^2 \cos nx = -n^2 y.$

(iii) $\dfrac{dy}{dx} = n e^{nx} \Rightarrow \dfrac{d^2 y}{dx^2} = n^2 e^{nx} = n^2 y$

(iv) $\dfrac{dy}{dx} = -n e^{-nx} \Rightarrow \dfrac{d^2 y}{dx^2} = n^2 e^{-nx} = n^2 y.$

(b) Functions (i) and (ii) are both solutions of the differential equation (7.46), but neither of them forms the general solution, which must contain two, as yet undetermined, constants of integration. Neither of functions (iii) nor (iv) are solutions because they lead to the constant n^2 rather than $-n^2$.

Neither of equations (iii) or (iv) are solutions to equation (7.46). However, if n was such that n^2 was negative, then both functions would be solutions to the equation. This would require us to define the square root of a negative number, which is at odds with our understanding of what constitutes a real number. In Chapter 2, Volume 2, we extend the concept of the number to include so-called imaginary and complex numbers, which embrace the idea that the square root of a negative number can be defined.

(c) We can obtain the general solution from the sum of constant multiples of the two solutions obtained in (a)(i) and (ii) above to give $y = A \cos nx + B \sin nx$. We can cross-check our answer by substituting the function and its second derivative into equation (7.46):

$$\frac{dy}{dx} = -nA \sin nx + nB \cos nx \Rightarrow \frac{d^2y}{dx^2} = -n^2(A \cos nx + B \sin nx)$$
$$= -n^2 y$$

(d) On substituting the boundary condition values $y = 0$, $x = 0$ into the general solution, we find that $0 = A$, and hence the solution becomes $y = B \sin nx$.

(e) We can further refine our solution by applying the second of our two boundary conditions. If $y = 0$ when $x = L$, then:

$$B \sin nL = 0$$

which is satisfied when $nL = m\pi$ (where m is an integer). Thus, $n = \frac{m\pi}{L}$ and the required solution is:

$$y = B \sin \left(\frac{m\pi}{L} x \right)$$

Note. This same solution with $m = 1$ arose in the context of the particle in the box problem in Problem 4.11.

Problem 7.7

(a) Use the results from part (a) of Worked Problem 7.4 to express the general solution of the differential equation $\frac{d^2y}{dx^2} = n^2 y$ in terms of exponential functions.

(b) Use the definitions for the hyperbolic functions, cosh x and sinh x, defined in Section 2.3.4, to show that the general solution obtained in (a) can be rewritten as:

$$y = (A + B) \cosh nx + (A - B) \sinh nx.$$

(c) Give the form of this solution when constrained by the boundary condition $y = 0$ when $x = 0$.

7.4.2 Second-order Differential Equations with Constant Coefficients

Linear second-order differential equations of the general form given in equation (7.45) are quite tricky to solve, but, fortunately, we are usually interested in situations where $P(x)$ and $S(x)$ are both constant functions, say c_1 and c_2:

$$\frac{d^2y}{dx^2} + c_1\frac{dy}{dx} + c_2y = Q(x) \qquad (7.47)$$

A number of possible variants of this equation can result from the different choices for c_1 and c_2:
(a) $c_1 = 0$ and $c_2 = 0$. This simple case results in the differential equation:

$$\frac{d^2y}{dx^2} = Q(x) \qquad (7.48)$$

Worked Problem 7.5

Q   Solve the second-order differential equation $\dfrac{d^2y}{dx^2} = 2x$.

A   The equation is solved by integrating once to give:

$$\frac{dy}{dx} = x^2 + C$$

and a second time to give:

$$y = \frac{x^3}{3} + Cx + D$$

---

which simply requires two steps of integration to yield an expression for y.
(b) $c_1 = 0$, $Q(x) = 0$ (the null function), c_2 positive. The differential equation now adopts the form of an eigenvalue problem (see equation 7.46):

$$\frac{d^2y}{dx^2} = -c_2y \qquad (7.49)$$

which is solved using the procedure described in Worked Problem 7.4. Likewise, for $c_2 < 0$ the solution is obtained using the same procedure.
(c) c_1 and c_2 are both positive and $Q(x) = 0$. Equation (7.47) now becomes a homogeneous linear second-order differential equation having the form:

$$\frac{d^2y}{dx^2} + c_1\frac{dy}{dx} = -c_2y \qquad (7.50)$$

Equation (7.50) is also an example of an eigenvalue problem (see Section 4.3.1), of the type commonly encountered in chemistry when modelling electronic and nuclear motions.

7.4.3 How is an Eigenvalue Problem Recognized?

The simplest way of thinking about an eigenvalue problem is to consider the result of some operator, $\hat{A}$, acting on a suitable function, $f(x)$, to yield a constant, λ, multiplied by the original function, $f(x)$:

$$\hat{A}f(x) = \lambda f(x) \tag{7.51}$$

The objective in this type of problem is to find the eigenfunctions, $f(x)$, and associated eigenvalues, λ, for a given operator, $\hat{A}$. The solution will generally yield a number of different eigenfunctions, and associated eigenvalues, all of which emerge from a single general solution.

The procedure used to solve second-order differential equations of the form of equation (7.50) is essentially the same as that described in Worked Problem 7.4 and involves the construction of trial eigenfunctions from some of the functions introduced in Chapter 2.

> The choice of $f(x)$ to label a function is entirely arbitrary: in principle, any label will do! We could just as easily have labelled it $\psi(x)$, $G(x)$, ϕ, γ or y. Later on, we do use the label y to represent the eigenfunction of a given operator, but y is also routinely used to name the value of the dependent variable for some given function, and as a result it can be confusing sometimes to distinguish between the label applied to the function itself and the label applied to the value of the function for a particular choice of independent variable.

The Search for Eigenfunctions

We first re-write equation (7.49), using differentiation operators $\hat{D}^2 = \dfrac{d^2}{dx^2}$ and $\hat{D} = \dfrac{d}{dx}$, to give:

$$\hat{D}^2 y + c_1 \hat{D} y = -c_2 y \tag{7.52}$$

Factorizing yields:

$$(\hat{D}^2 + c_1 \hat{D})y = -c_2 y \tag{7.53}$$

which is of the form of an eigenvalue equation:

$$\hat{A}y = -c_2 y,$$

where:

$$\hat{A} = \hat{D}^2 + c_1 \hat{D} \tag{7.54}$$

is the operator; y is the eigenfunction and $-c_2$ the eigenvalue. The form of the operator $\hat{A}$ is such that its eigenfunctions must be functions whose first and second derivatives differ only by a constant.

Worked Problem 7.6

Q Which of the functions $y = \cos nx$, $y = \sin nx$, $y = e^{nx}$ and $y = e^{-nx}$ are eigenfunctions of: (a) $\hat{D}$; (b) $\hat{D}^2$; (c) $\hat{D}^2 + \hat{D}$?
For each eigenfunction, give the associated eigenvalue.

A (a) $\hat{D} \cos nx = -n \sin nx$; $\hat{D} \sin nx = n \cos nx$; $\hat{D} e^{nx} = n e^{nx}$;
$\hat{D} e^{-nx} = -n e^{-nx}$. Only the two exponential functions are eigenfunctions of $\hat{D}$, with associated eigenvalues n and $-n$, respectively.
(b) $\hat{D}^2 \cos nx = -n^2 \cos nx$; $\hat{D}^2 \sin nx = -n^2 \sin nx$; $\hat{D}^2 e^{nx} = n^2 e^{nx}$;
$\hat{D}^2 e^{-nx} = n^2 e^{-nx}$. All four functions are eigenfunctions of $\hat{D}^2$, with respective eigenvalues of $-n^2$, $-n^2$, n^2 and n^2.
(c) $(\hat{D}^2 + \hat{D}) \cos nx = -n^2 \cos nx - n \sin nx$; $(\hat{D}^2 + \hat{D}) \sin nx = -n^2 \sin nx + n \cos nx$; $(\hat{D}^2 + \hat{D}) e^{nx} = n^2 e^{nx} + n e^{nx} = (n^2 + n) e^{nx}$; $(\hat{D}^2 + \hat{D}) e^{-nx} = n^2 e^{-nx} - n e^{-nx} = (n^2 - n) e^{-nx}$. Only the two exponential functions are eigenfunctions of both operators, with eigenfunctions $n^2 + n$ and $n^2 - n$, respectively.

The key feature emerging from Worked Problem 7.6 is that if two functions, f_1 and f_2, are eigenfunctions of an operator $\hat{A}$, and have the same eigenvalues, λ, then an arbitrary linear combination of the two functions is also an eigenfunction of $\hat{A}$ with eigenvalue λ. In this example, the two functions $y = \cos nx$ and $y = \sin nx$ are both eigenfunctions of the $\hat{D}^2$ operator with eigenvalue $-n^2$. Consequently, it follows that an arbitrary linear combination of the two functions is also an eigenfunction of this operator with eigenvalue $-n^2$. This concept is expressed formally in the following expression:

$$\hat{A}(bf_1 + cf_2) = b\lambda f_1 + c\lambda f_2 = \lambda(bf_1 + cf_2) \qquad (7.55)$$

Problem 7.8

Show that the function $y = A \cos nx + B \sin nx$ is an eigenfunction of the $\hat{D}^2$ operator with eigenvalue $-n^2$.

Worked Problem 7.7

Q The differential equation $\dfrac{d^2 y}{dx^2} - 3\dfrac{dy}{dx} + 2y = 0$ has the same form as equation (7.50) and, as such, can be written as an eigenvalue equation, $\hat{A}y = \lambda y$.

(a) Give the form of the operator $\hat{A}$, expressed in terms of the $\hat{D}^2$ and $\hat{D}$ operators, and the value of λ, the eigenvalue.

(b) Find the eigenfunctions of $\hat{A}$, and hence deduce the general solution.

(c) Give the form of the solution for the boundary conditions $y = 0$ when $x = 0$ and $y = e$ when $x = 1$.

A (a) The operator $\hat{A} = D^2 - 3\hat{D}$ and has the same form as that given in equation (7.54) with $c_1 = -3$. The eigenvalue $\lambda = -2$.

(b) We have seen already in Worked Problem 7.6 that e^{nx} is an eigenfunction of the operators $\hat{D}^2$ and $\hat{D}$. It follows that e^{nx} is also an eigenfunction of the operator $\hat{D}^2 + c_1\hat{D}$ and hence of $\hat{A} = \hat{D}^2 - 3\hat{D}$. We can now tailor this function by finding the appropriate values of n that are consistent with this operator, and an eigenvalue $\lambda = -2$. Thus, since:

$$\hat{D}^2 e^{nx} = n^2 e^{nx} \text{ and } -3\hat{D}e^{nx} = -3ne^{nx}$$

it follows that:

$$\hat{A}e^{nx} = (n^2 - 3n)e^{nx} = -2e^{nx}$$

and, on cancelling the e^{nx} terms, we obtain $(n^2 - 3n) = -2$, which may be written as:

$$n^2 - 3n + 2 = 0 \tag{7.56}$$

The two roots of the quadratic equation (7.56) are $n = 2$ and $n = 1$, and it follows that e^{2x} and e^x are two solutions of the differential equation which we may use to build the general solution in the form $y = Be^{2x} + Ce^x$.

(c) If we now apply the two boundary conditions $y = 0$, $x = 0$ and $y = e$, $x = 1$, we obtain the two equations $B + C = 0$ and $Be^2 + Ce = e$. Solving for A and B yields:

$$C = \frac{1}{1 - e} \text{ and } B = -\frac{1}{1 - e}$$

and the final solution then takes the form:

$$y = -\frac{1}{e - 1}e^{2x} + \frac{1}{e - 1}e^x$$

$$\Rightarrow y = \frac{1}{e - 1}\left(e^x - e^{2x}\right)$$

Problem 7.9

Solve the differential equation $\dfrac{d^2y}{dx^2} - 5\dfrac{dy}{dx} + 6y = 0$, subject to the boundary conditions $y = 0$ when $x = 0$ and $dy/dx = 1$ when $x = 0$.

Summary of Key Points

This chapter has been concerned with bringing differential and integral calculus together, in order to solve a number of differential equations that are used widely in chemistry. The key points discussed in this chapter include:

1. The order of a differential equation.

2. Creating a differential equation using the first or second derivative of a function.

3. Examples of differential equations in a chemical context.

4. The solution to first-order differential equations of the form $\dfrac{dy}{dx} = f(x)g(y)$ using the separation of variables method.

5. The application of boundary conditions to determine the value of constant(s) of integration.

6. Finding general solutions to linear first order differential equations using the integrating factor method.

7. Finding general solutions to linear second-order differential equations by substitution of trial functions.

8. A revision of the eigenvalue problem and the construction of trial functions to provide a solution of second-order differential equations with constant coefficients.

Answers to Problems

Chapter 1

1.1. (a)   (i)   2.554 455 = 2.554 46 to 5 d.p.
                 2.554 5 to 4 d.p.
                 2.554 to 3 d.p.
      (ii)   2.554 455 = 2.55 to 3 sig. fig.

   (b)   (i)   1.723 205 08 = 1.723 21 to 5 d.p.
                     = 1.723 2 to 4 d.p.
                     = 1.723 to 3 d.p.
      (ii)   1.723 205 08 = 1.72 to 3 sig. fig.

   (c)   (i)   3.141 592 6 . . . = 3.141 59 to 5 d.p.
                      = 3.141 6 to 4 d.p.
                      = 3.142 to 3 d.p.
      (ii)   3.141 592 6 . . . = 3.14 to 3 sig. fig.

   (d)   (i)   2.718 281 828 = 2.718 28 to 5 d.p.
                      = 2.718 3 to 4 d.p.
                      = 2.718 to 3 d.p.
      (ii)   2.718 281 2 . . . = 2.72 to 3 sig. fig.

1.2. (a)   (i)   1.378 423 784 2 rational $\Rightarrow$ 1.378 to 4 sig. fig.
      (ii)   1.378 423 784 2 . . . irrational $\Rightarrow$ 1.378 to 4 sig. fig.
      (iii)   1/70 = 0.014 285 714 . . . rational $\Rightarrow$ 0.014 29 to 4 sig. fig.
      (iv) $\pi/4$ = 0.785 398 163 . . . irrational $\Rightarrow$ 0.785 4 to 4 sig. fig.
      (v)   0.005068 rational $\Rightarrow$ 0.005068 to 4 sig. fig.

(vi) $e/10 = 0.271\ 828\ 182...$ irrational $\Rightarrow 0.271\ 8$ to 4 sig. fig.

(b)   23.3 cm^3: 3 sig. figs; 1 d.p. Max. titre 23.35 cm^3; min. titre 23.25 cm^3.

1.3. (a) $\dfrac{10^2 \times 10^{-4}}{10^6} = \dfrac{10^{-2}}{10^6} = 10^{-8}$.

(b) $\dfrac{9 \times 2^4 \times 3^{-2}}{4^2} = \dfrac{2^4}{4^2} = \dfrac{2^4}{(2^2)^2} = \dfrac{2^4}{2^4} = 1$.

(c) $\left(\dfrac{10}{3^2 + 4^2 + 5^2}\right)^{-1/2} = \left(\dfrac{10}{9 + 16 + 25}\right)^{-1/2} = \left(\dfrac{10}{50}\right)^{-1/2} = \left(\dfrac{1}{5}\right)^{-1/2}$

$= (5^{-1})^{-1/2} = 5^{1/2}$

(d) $\dfrac{(2^4)^3}{4^4} = \dfrac{2^{12}}{4^4} = \dfrac{2^{12}}{(2^2)^4} = \dfrac{2^{12}}{2^8} = 2^4 = 16$.

1.4. (a) $(2.5 \times 10^2 - 0.5 \times 10^2)^2 / 4 \times 10^4 = (250 - 50)^2 / 4 \times 10^4$
$= 200^2 / 40000 = 1$.

(b) $\left(\dfrac{1}{2 \times 4}\right)^{1/3} - 4 \times \dfrac{1}{16} = \dfrac{1}{8^{1/3}} - \dfrac{1}{4} = \dfrac{1}{2} - \dfrac{1}{4} = \dfrac{1}{4}$.

1.5. (a) $IE = \dfrac{2.179\ aJ}{0.1602\ aJ\ eV^{-1}} = 13.6$ eV to 3 sig. fig.

(b) $\varepsilon_{vib} = 6.626 \times 10^{-34}\ Js \times 1.2404 \times 10^{14}\ s^{-1} = 8.22 \times 10^{-20}$ J.

1.6. 1×10^6 cm^3 = 1 m^3;   therefore $\rho = \dfrac{879\ kg\ m^{-1}}{1 \times 10^6\ cm^3\ m^{-1}} = 8.79 \times 10^{-4}$ kg cm^{-3} and so 1 cm^3 of benzene weighs 8.79×10^{-4} kg, which is equivalent to

$$\dfrac{8.79 \times 10^{-4}\ kg}{0.078\ kg\ mol^{-1}} = 0.0113\ mol.$$

1.7. Benzene diameter = 600 pm = 600×10^{-12} m; circumference of the Earth = $2\pi \times 6.378 \times 10^6$ m; therefore:

$$n = \dfrac{2\pi \times 6.378 \times 10^6\ m}{600 \times 10^{-12}\ m} = 6.679 \times 10^{16} \quad \Rightarrow \quad \dfrac{6.679 \times 10^{16}}{6.022 \times 10^{23}}$$
$$= 1.109 \times 10^{-7}\ moles.$$

1.109×10^{-7} moles of benzene weighs:

$1.109 \times 10^{-7} \times 0.078$ kg $= 8.65 \times 10^{-9}$ kg $= 8.65 \times 10^{-6}$ g $= 8.65$ μg

1.8. Volume of cube $= a^3 = 6.7832 \times 10^{-29} \text{m}^3$, so mass of Au $=$ $19.321 \times 10^3 \text{k g m}^{-3} \times 6.7832 \times 10^{-29} \text{k g} = 1.3106 \times 10^{-24}$ kg $=$ 1.3106×10^{-21} g. RMM ^{197}Au$(100\%) = 196.97 \text{ g mol}^{-1}$ so 1 molecule of Au $= 3.2708 \times 10^{-22}$ g and therefore $n = \dfrac{1.3106 \times 10^{-21}}{3.2708 \times 10^{-22}} = 4.007$.

1.9. (a) $2 < 6; 6 > 2$.
(b) $1.467 < 1.469; 1.469 > 1.467$.
(c) $\pi > e; e < \pi$.

1.10. (a) $2.4555 < 2.456 < 2.4565$.
(b) $-5.34 > -5.35; 5.34 < 5.35$.
     $5.34 > -5.35; -5.34 < 5.35$.

1.11. (a) $|4 - 9| = 5; |-3 - 6| = 9; |9 - 4| = 5$.
(b) $0, \infty, 0, -\infty$.

1.12. (a) $S_{100} = \dfrac{100 \times 101}{2} = 5050$.

(b) $S_{68} = \dfrac{-68 \times 69}{2} = -2346$.

1.13. (a) 3 spin states, 2 nuclei $\Rightarrow 3^2 = 9$ spin states.
         3 spin states, 3 nuclei $\Rightarrow 3^3 = 27$ spin states.
(b) Number of spin states associated with n equivalent nuclei with spin $I = (2I + 1)^n$.
(c) ^{51}V $\Rightarrow I = 7/2$, so for a single atom of ^{51}V there are $(2 \times 7/2) + 1 = 8$ spin states.

1.14. (a) $\dfrac{(u^2 + v^2)}{(v - u)} = \dfrac{2(x^2 + y^2)}{-2y} = \dfrac{-(x^2 + y^2)}{y}$.

(b) $\dfrac{uv}{2u - v} = \dfrac{x^2 - y^2}{x + 3y}$.

(c) $\dfrac{10^{u+v}}{10^{u-v}} = \dfrac{10^{2x}}{10^{2y}}$.

1.15. (a) (i) $4p - q - (2q + 3p) = p - 3q$.
          (ii) $3p^2 - p(4p - 7) = -p^2 + 7p$.
(b) (i) $(1 + x)^2 - (1 - x)^2 = 4x$.
     (ii) $x(2x + 1) - (1 + x - x^2) = 3x^2 - 1$.

1.16. (a) $\dfrac{p^4 q^2}{p^2 q^3} = \dfrac{p^2}{q}; \, q \neq 0, p \neq 0.$

(b) $\dfrac{p^8 q^{-3}}{p^{-5} q^2} = \dfrac{p^{13}}{q^5}; \, q \neq 0, p \neq 0.$

(c) $\dfrac{4x}{6x^2 - 2x} = \dfrac{4}{6x - 2} = \dfrac{2}{3x - 1}; \, x \neq \dfrac{1}{3}, 0.$

(d) $\dfrac{3x^2 - 12xy}{3} = x^2 - 4xy.$

1.17. (a) (i) $x^2 - 3x + 2 = (x - 1)(x - 2)$, since the solutions to $a^2 - 3a + 2 = 0$ are $a = 1, 2.$

(ii) $x^3 - 7x + 6 = (x - 1)(x - 2)(x + 3)$, since the solutions to $a^3 - 7a + 6 = 0$ are $a = 1, \, 2, \, -3.$

(b) (i) $\dfrac{x^3 - 7x + 6}{x - 2} = \dfrac{(x - 1)(x - 2)(x + 3)}{x - 2} = (x - 1)(x + 3).$

(ii) $\dfrac{x^2 - 1}{x - 1} = \dfrac{(x + 1)(x - 1)}{x - 1} = x + 1.$

1.18. (a) (i) $\dfrac{3x}{4} - \dfrac{x}{2} = \dfrac{3x - 2x}{4} = \dfrac{x}{4}.$

(ii) $\dfrac{2}{x} - \dfrac{1}{x^2} = \dfrac{2x - 1}{x^2}.$

(iii) $1 - \dfrac{1}{x} + \dfrac{2}{x^2} = \dfrac{x^2 - x + 2}{x^2}.$

(b) (i) $\dfrac{1}{1 + x} - \dfrac{1}{1 - x} = \dfrac{1 - x - 1 - x}{(1 + x)(1 - x)} = \dfrac{-2x}{1 - x^2} = \dfrac{2x}{x^2 - 1}.$

(ii) $\dfrac{2x}{x^2 + 1} - \dfrac{2}{x} = \dfrac{2x^2 - 2x^2 - 2}{x(x^2 + 1)} = \dfrac{-2}{x(x^2 + 1)}.$

1.19. (a) $\dfrac{RT}{F} \Rightarrow \dfrac{\text{J K}^{-1} \, \text{mol}^{-1} \times \text{K}}{\text{C mol}^{-1}} = \text{J C}^{-1} = \text{C V C}^{-1} = \text{V}.$

(b) $\dfrac{m_e e^4}{8h^3 c \varepsilon_0^2} \Rightarrow \dfrac{\text{kg C}^4}{\text{J}^3 \, \text{s}^3 \, \text{m s}^{-1} \, \text{J}^{-2} \, \text{C}^4 \, \text{m}^{-2}} = \dfrac{\text{kg}}{\text{J s}^2 \, \text{m}^{-1}}$

$= \text{kg J}^{-1} \, \text{m s}^{-2} = \text{m}^{-1}$, since $1 \, \text{J} = 1 \, \text{kg m}^2 \, \text{s}^{-2}.$

Chapter 2

2.1. (a) Yes, domain registration numbers.

(b) No. One keeper may own > 1 car.

(c) No. More than 1 element per group.

(d) Yes. Each element only belongs to one group.

2.2. $s = \dfrac{ezE}{6\pi\eta a} \Rightarrow \dfrac{\text{C V m}^{-1}}{\text{m}^{-1}\,\text{kg s}^{-1}\,\text{m}} = \text{C V m}^{-1}\,\text{kg}^{-1}\,\text{s}$

$\quad\quad = \text{J m}^{-1}\,\text{kg}^{-1}\,\text{s} = \text{kg m}^2\,\text{s}^{-2}\,\text{m}^{-1}\,\text{kg}^{-1}\,\text{s} = \text{m s}^{-1}.$

2.3. (a) $f(x) = \begin{cases} x - 1, & x \geqslant 1 \\ -x + 1, & x < 1 \end{cases}.$

(b)

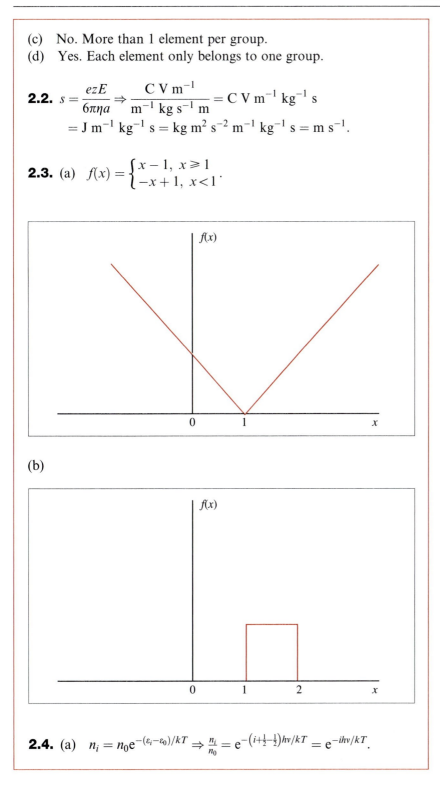

2.4. (a) $n_i = n_0 e^{-(\varepsilon_i - \varepsilon_0)/kT} \Rightarrow \dfrac{n_i}{n_0} = e^{-\left(i + \frac{1}{2} - \frac{1}{2}\right)h\nu/kT} = e^{-ih\nu/kT}.$

(b)

i	1	2	3	4	5
n_i/n_0	1.583×10^{-3}	2.506×10^{-6}	3.966×10^{-9}	6.278×10^{-12}	9.938×10^{-15}

2.5. (a) (i) $\log 4 = \log 2^2 = 2\log 2$. (ii) $\log 8 = \log 2^3 = 3\log 2$.
(iii) $\log 6 - \log 3 = \log \frac{6}{3} = \log 2$. (iv) $\ln 8 = 3\ln 2 = 3\ln 10 \log 2$.
(v) $\ln \frac{1}{2} = \ln 2^{-1} = -\ln 2$.
(b) (i) $\log 2 + \log 3 = \log 2 \times 3 = \log 6$. (ii) $\ln 3 - \ln 6 = \log \frac{3}{6} = \log \frac{1}{2}$.

2.6. (a) Given that $pH = -\log a_H$ and $\log_{10} y = \log_e y \log_{10} e$, it follows that $\log a_H = \ln a_H \log e$ and so $pH = -\log e \ln a_H$.

(b) $E^{\ominus} = -\frac{RT}{nF}\ln K$ and $\ln K = \frac{1}{\log e}\log K$, so $E^{\ominus} = -2.303 \frac{RT}{nF}\log K$.
(c) $pK = 4.756 = -\log K \Rightarrow K = 10^{-4.756} = 1.754 \times 10^{-5}$.

2.7. (a) $r_{H-H} = 2 \times r_{P-H} \times \sin(\theta/2) = 2 \times 140 \times \sin 61.5°$
$$= 246.1 \text{ pm.}$$
(b) $(r_{H-H})_{\text{excited state}} = 195.1$ pm, and so $\Delta r_{H-H} = 51$ pm.

2.8. $\cot \theta = \frac{1}{\tan \theta} = \frac{\cos \theta}{\sin \theta}$; domain all real numbers except those values of θ for which $\sin \theta = 0$, *i.e.* $\theta = n\pi$, $n = 0, \pm 1, \pm 2, \pm 3, \dots$ Similarly, $\text{cosec } \theta = \frac{1}{\sin \theta}$ has the same domain as $\cot \theta$.

2.9. $\cos(-\theta) = \cos(-\theta + 2\pi) \Rightarrow \cos(-\theta) = \cos 2\pi \cos \theta + \sin 2\pi \sin \theta = \cos \theta$.
$$\tan(-\theta) = \frac{\sin(-\theta)}{\cos(-\theta)} = \frac{-\sin \theta}{\cos \theta} = -\tan\theta.$$

2.10. (a) (i) $\sinh x + \cosh x = \frac{1}{2}e^x - \frac{1}{2}e^{-x} + \frac{1}{2}e^x + \frac{1}{2}e^{-x} = e^x$.
(ii) $\sinh x - \cosh x = \frac{1}{2}e^x - \frac{1}{2}e^{-x} - \frac{1}{2}e^x - \frac{1}{2}e^{-x} = -e^{-x}$.

(b) (i) $\cosh^2 x - \sinh^2 x = \frac{1}{4}e^{2x} + \frac{1}{4} + \frac{1}{4} + \frac{1}{4}e^{-2x} - \frac{1}{4}e^{2x} + \frac{1}{4} + \frac{1}{4}$
$-\frac{1}{4}e^{-2x} = 1$.
(ii) $\cosh^2 x + \sinh^2 x = \frac{1}{4}e^{2x} + \frac{1}{4} + \frac{1}{4} + \frac{1}{4}e^{-2x} + \frac{1}{4}e^{2x}$
$$-\frac{1}{4} - \frac{1}{4} + \frac{1}{4}e^{-2x} = \frac{1}{2}e^{2x} + \frac{1}{2}e^{-2x} = \cosh 2x.$$

(iii) $\sinh x \cosh x = (\tfrac{1}{2}e^x - \tfrac{1}{2}e^{-x})(\tfrac{1}{2}e^x + \tfrac{1}{2}e^{-x})$

$$= \tfrac{1}{4}e^{2x} + \tfrac{1}{4} - \tfrac{1}{4} - \tfrac{1}{4}e^{-2x} = \tfrac{1}{4}(e^{2x} - e^{-2x})$$

$$= \tfrac{1}{2}\sinh 2x \quad \Rightarrow \quad \sinh 2x = 2 \sinh x \cosh x.$$

2.11. (a) (i) $x^2 + x - 6 = 0 \Rightarrow \dfrac{-1 \pm \sqrt{1^2 - 4.1.-6}}{2} = -\dfrac{1}{2} \pm \dfrac{\sqrt{25}}{2} = 2, -3.$

(ii) $x^2 - 1 = 0 \Rightarrow \dfrac{0 \pm \sqrt{0^2 - 4.1.-1}}{2} = \pm \dfrac{\sqrt{4}}{2} = \pm 1.$

(iii) $x^2 - 2\sqrt{2}x + 2 = 0 \Rightarrow \dfrac{2\sqrt{2} \pm \sqrt{8 - 4.1.2}}{2} = \sqrt{2}$ twice.

(b) (i) $x^2 + x - 6 = (x - 2)(x + 3).$
(ii) $x^2 - 1 = (x + 1)(x - 1).$
(iii) $x^2 - 2\sqrt{2}x + 2 = (x - \sqrt{2})(x - \sqrt{2}).$

2.12. (a) If $R_{3s} = N\left\{27 - \dfrac{18r}{a_0} + 2\dfrac{r^2}{a_0^2}\right\}e^{-\frac{r}{3a_0}}$, then $R_{3s,r=0} = 27N.$

$\lim\limits_{r \to \infty} R_{3s} = N\{27 - \text{v. large} + \text{v. v. large}\} \times \text{v. v. v. small} = 0.$

(b) $\left\{27 - 18\left(\dfrac{r}{a_0}\right) + 2\left(\dfrac{r}{a_0}\right)^2\right\} = 0$ when

$$\dfrac{r}{a_0} = \dfrac{18}{4} \pm \dfrac{\sqrt{18^2 - 4.27.2}}{4} = \dfrac{9}{2} \pm \dfrac{\sqrt{108}}{4} = \dfrac{9}{2} \pm \dfrac{5.196}{2}$$

$$\Rightarrow \dfrac{r}{a_0} = 7.098 \text{ and } 1.902 \text{ or } r = 7.098a_0 \text{ and } 1.902a_0.$$

(c)

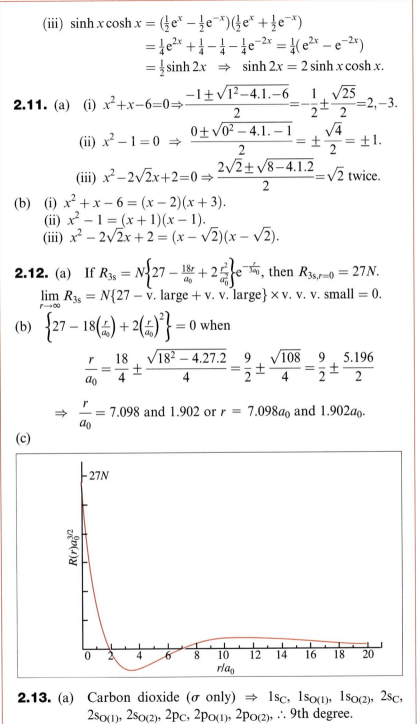

2.13. (a) Carbon dioxide (σ only) $\Rightarrow$ $1s_C$, $1s_{O(1)}$, $1s_{O(2)}$, $2s_C$, $2s_{O(1)}$, $2s_{O(2)}$, $2p_C$, $2p_{O(1)}$, $2p_{O(2)}$, $\therefore$ 9th degree.
(b) Benzene (π only) $\Rightarrow$ $6 \times 2p_{C(\pi)}$, $\therefore$ 6th degree.

2.14. (a) $K = hc/e$; $K_w = hb$

(c) $e = 1 - c$

(d) $K = \frac{hc}{e} \Rightarrow K = \frac{hc}{1-c} \Rightarrow (1-c)K = hc \Rightarrow \frac{1-c}{c} = \frac{h}{K} \Rightarrow \frac{1}{c} - 1$

$= \frac{h}{K} \Rightarrow \frac{1}{c} = \frac{h}{K} + 1$ and so $c = \frac{1}{h/K+1}$.

(e) $\frac{1}{h/K+1} + b = 1 + h \Rightarrow b = 1 + h - \frac{1}{h/K+1} = 1 + h - \frac{K}{h+K}$.

(f) $K_w = hb = h + h^2 - \frac{Kh}{h+K}$.

(g) $K_w = h\frac{K+h}{K+h} + h^2\frac{K+h}{K+h} - \frac{Kh}{h+K}$

$= \frac{1}{K+h}\left(hK + h^2 + h^2K + h^3 - Kh\right)$

$\Rightarrow K_w = \frac{1}{K+h}\left(h^3 + h^2(1+K) + hK - hK\right)$

$= \frac{1}{K+h}\left(h^3 + h^2(1+K)\right)$

$\Rightarrow K_wK + K_wh = h^3 + h^2(1+K)$

$\Rightarrow h^3 + h^2(1+K) - K_wK - K_wh = 0.$

(h) $h^3 + 1.000018h^2 - 10^{-14}h - 1.8 \times 10^{-19}$

(i) $h = 4.243 \times 10^{-9}$ and $c = \frac{1}{h/K+1}$, so

$c = \frac{1}{(4.243 \times 10^{-9}/1.8 \times 10^{-5}) + 1} = 0.999764333$

$e = 1 - c = 1 - 0.999764333 = 2.3567 \times 10^{-4}$.

$c + b = 1 + h \Rightarrow$

$b = 1 + h - c = 1 + 4.243 \times 10^{-9}$

$- 0.999764333 = 2.3567 \times 10^{-4}$.

Chapter 3

3.1. (a) $\lim_{x \to \infty} x^2 e^{-x} = 0$; the e^{-x} term dominates.

(b) $\lim_{x \to \infty} \cos(2x)e^{-x} = 0$; again the e^{-x} term dominates.

3.2. (a) $\lim_{x \to 0} x^2 e^{-x} = 0$; the x^2 term dominates as $\lim_{x \to 0} e^{-x} = 1$.

(b) $\lim_{x \to 0} \cos(2x)e^{-x} = 1$, as both $\cos(2x)$ and e^{-x} tend to 1 as $x \to 0$.

3.3. (a) $f(x) = \dfrac{2x}{x-4}$ is indeterminate at $x = 4$. Thus

$$\lim_{x \to 4} \frac{2x}{x-4} \equiv \lim_{\delta \to 0} \frac{2(4+\delta)}{4+\delta-4} = \frac{8}{\delta} = \infty.$$

(b) $f(x) = \dfrac{x^2 - 4}{x+2}$ is indeterminate at $x = -2$. Thus

$$\lim_{x \to -2} \frac{x^2-4}{x+2} \equiv \lim_{\delta \to 0} \frac{(-2+\delta)^2 - 4}{-2+\delta+2} = \lim_{\delta \to 0} \frac{4 - 4\delta + \delta^2 - 4}{\delta}$$
$$= \lim_{\delta \to 0} -4 + \delta = -4.$$

(c) $f(x) = \dfrac{x-1}{x^2-1}$ is indeterminate at $x = 1, -1$; Thus

$$\lim_{x \to +1} \frac{x-1}{x^2-1} \equiv \lim_{\delta \to 0} \frac{1+\delta-1}{(1+\delta)^2 - 1} = \lim_{\delta \to 0} \frac{\delta}{2\delta - \delta^2} = \lim_{\delta \to 0} \frac{1}{2-\delta} = \frac{1}{2}.$$

$$\lim_{x \to -1} \frac{x-1}{x^2-1} \equiv \lim_{\delta \to 0} \frac{-1+\delta-1}{(-1+\delta)^2 - 1} = \lim_{\delta \to 0} \frac{-2+\delta}{-2\delta + \delta^2} = \lim_{\delta \to 0} \frac{-2}{-2\delta} = \infty.$$

(d) $f(x) = 3x^2 - \dfrac{2}{x} - 1$ is indeterminate at $x = 0$. Thus

$$\lim_{x \to 0} (3x^2 - \frac{2}{x} - 1) = -\infty.$$

3.4. (a) $\lim_{x \to \infty} \dfrac{5}{x+1} = 0.$

(b) $\lim_{x \to \infty} \dfrac{3x}{x-4} = \dfrac{3x}{x} = 3.$

(c) $\lim_{x \to \infty} \dfrac{x^2}{x+1} = \dfrac{x^2}{x} = x = \infty.$

(d) $\lim_{x \to \infty} \dfrac{x+1}{x+2} = \dfrac{x}{x} = 1.$

3.5. $\lim_{x \to 0} (\ln x - \ln 2x) = \lim_{x \to 0} \left(\ln \dfrac{x}{2x} \right) = \ln \dfrac{1}{2}.$

3.6. $C_V = 3R(ax)^2 \left\{ \dfrac{e^{ax/2}}{e^{ax} - 1} \right\}^2$

$$\lim_{x \to \infty} 3R(ax)^2 \left\{ \frac{e^{ax/2}}{e^{ax} - 1} \right\}^2 = 3R(ax)^2 \left\{ e^{-ax/2} \right\}^2 = 3R(ax)^2 e^{-ax} = 0 \quad \text{since}$$

the exponential term will dominate. Therefore $C_V \to 0$ as $T \to 0\,\mathrm{K}$.

3.7. (a) $\lim_{r \to 0} R_{3s} = N \left(\dfrac{r}{a_0} \right)^2 \times 1 = 0.$

(b) $\lim_{r \to \infty} R_{3s} = N \left(\dfrac{r}{a_0} \right)^2 \times 0 = 0.$

3.8. (a)   For $k_{-1} \gg k_2[H_2]$ $k_{-1} + k_2[H_2] \approx k_{-1}$ and so

$$\frac{d[N_2]}{dt} = \frac{k_1 k_2 [H_2][NO]^2}{k_{-1}}.$$

(b)   For $k_{-1} \ll k_2[H_2]$, $k_{-1} + k_2[H_2] \approx k_2[H_2]$ and so

$$\frac{d[N_2]}{dt} = \frac{k_1 k_2 [H_2][NO]^2}{k_2 [H_2]} = k_1 [NO]^2.$$

Chapter 4

4.1. $\dfrac{dy}{dx} = \lim\limits_{\Delta x \to 0} \left\{ \dfrac{f(x + \Delta x) - f(x)}{\Delta x} \right\} = \lim\limits_{\Delta x \to 0} \left\{ \dfrac{3 - 3}{\Delta x} \right\} = 0.$

4.2. (a) $\dfrac{dy}{dx} = \lim\limits_{\Delta x \to 0} \left\{ \dfrac{3(x+\Delta x)^2 - 3x^2}{\Delta x} \right\} = \lim\limits_{\Delta x \to 0} \left\{ \dfrac{3x^2 + 6x\Delta x + 3\Delta x^2 - 3x^2}{\Delta x} \right\}$

$$= \lim\limits_{\Delta x \to 0} 6x + \Delta x = 6x$$

(b) $\dfrac{dy}{dx} = \lim\limits_{\Delta x \to 0} \left\{ \dfrac{\frac{1}{(x+\Delta x)^2} - \frac{1}{x^2}}{\Delta x} \right\} = \lim\limits_{\Delta x \to 0} \left\{ \dfrac{x^2 - (x + \Delta x)^2}{x^2(x + \Delta x)^2 \Delta x} \right\}$

$$= \lim\limits_{\Delta x \to 0} \left\{ \dfrac{x^2 - x^2 - 2x\Delta x - \Delta x^2}{x^2(x + \Delta x)^2 \Delta x} \right\}$$

$$= \lim\limits_{\Delta x \to 0} \left\{ \dfrac{-2x - \Delta x}{x^2(x + \Delta x)^2} \right\} = -\dfrac{2x}{x^4} = -\dfrac{2}{x^3}.$$

4.3. (a) $\dfrac{d}{dx} x^{3/4} = \dfrac{3}{4} x^{-1/4}.$ (b) $\dfrac{d}{dx} e^{-3x} = -3e^{-3x}.$

(c) $\dfrac{d}{dx} 1/x = -x^{-2}.$ (d) $\dfrac{d}{dx} a\cos ax = -a^2 \sin ax.$

4.4. $\left(\dfrac{d}{dx} + 2 \right) e^{-2x} = \dfrac{d}{dx} e^{-2x} + 2e^{-2x} = -2e^{-2x} + 2e^{-2x} = 0.$

4.5. (a) $\dfrac{d}{dx}(x - 1)(x^2 + 4) = 3x^2 - 2x + 4.$

(b) $\dfrac{d}{dx} \dfrac{x}{x + 1} = \dfrac{1}{(x + 1)^2}.$

(c) $\dfrac{d}{dx} \sin^2 x = 2\sin x \cos x = \sin 2x.$

(d) $\dfrac{d}{dx} x \ln x = 1 + \ln x.$

(e) $\dfrac{d}{dx}e^x\sin x = e^x(\cos x + \sin x).$

4.6. $y = e^{x\sin x}; u = x\sin x \Rightarrow y = e^u$

$\dfrac{dy}{du} = e^u ; \dfrac{du}{dx} = x\cos x + \sin x \Rightarrow$

$\dfrac{dy}{dx} = \dfrac{dy}{du}\times\dfrac{du}{dx} = e^u(x\cos x + \sin x) = e^{x\sin x}(x\cos x + \sin x).$

4.7. (a) $y = \ln(2 + x^2); u = 2 + x^2 \Rightarrow y = \ln u$

$\dfrac{dy}{du} = \dfrac{1}{u}; \dfrac{du}{dx} = 2x \Rightarrow \dfrac{dy}{dx} = \dfrac{1}{u}\times 2x = \dfrac{2x}{2 + x^2}.$

(b) $y = 2\sin(x^2 - 1); u = x^2 - 1 \Rightarrow y = 2\sin u$

$\dfrac{dy}{du} = 2\cos u; \dfrac{du}{dx} = 2x \Rightarrow \dfrac{dy}{dx} = 2\cos u \times 2x = 4x\cos(x^2 - 1).$

4.8. (a) $y = 1/x; \dfrac{dy}{dx} = -x^{-2}; \dfrac{d^2y}{dx^2} = 2x^{-3}; \dfrac{d^3y}{dx^3} = -6x^{-4}.$

(b) $y = N\sin ax; \dfrac{dy}{dx} = Na\cos ax; \dfrac{d^2y}{dx^2} = -Na^2\sin ax;$

$\dfrac{d^3y}{dx^3} = -Na^3\cos ax.$

4.9. (a) $\dfrac{d^2}{dx^2}x^3 = 6x.$ (b) $\dfrac{d^2}{dx^2}\sin kx = -k^2\sin kx.$

(c) $\dfrac{d^2}{dx^2}\{\sin kx + \cos kx\} = -k^2\{\sin kx + \cos kx\}.$

(d) $\dfrac{d}{dx}e^{ax} = ae^{ax}.$

4.10. $\left(\dfrac{d^2}{dx^2} - 2\dfrac{d}{dx} - 3\right)e^{mx} = m^2 e^{mx} - 2me^{mx} - 3e^{mx}$

$= (m^2 - 2m - 3)e^{mx}.$

$\hat{A}$ annihilates $f(x)$ when $m^2 - 2m - 3 = 0,$ *i.e.*
when $(m + 1)(m - 3) = 0.$
$\Rightarrow m = -1, 3.$

4.11. $-\dfrac{h^2}{8\pi^2 m}\dfrac{d^2\psi}{dx^2} = E\psi; \psi = \sqrt{\dfrac{2}{L}}\sin\dfrac{\pi x}{L}$

$\Rightarrow -\dfrac{h^2}{8\pi^2 m}\dfrac{d^2}{dx^2}\left(\sqrt{\dfrac{2}{L}}\sin\dfrac{\pi x}{L}\right) = E\sqrt{\dfrac{2}{L}}\sin\dfrac{\pi x}{L}$

$\Rightarrow \dfrac{h^2}{8\pi^2 m}\sqrt{\dfrac{2}{L}}\left(\dfrac{\pi}{L}\right)^2\sin\dfrac{\pi x}{L} = E\sqrt{\dfrac{2}{L}}\sin\dfrac{\pi x}{L}$

$$\Rightarrow \frac{h^2}{8\pi^2 m}\left(\frac{\pi}{L}\right)^2 = E$$

$$\Rightarrow E = \frac{h^2}{8mL^2}.$$

4.12. (a) $D(r) = Nr^2 e^{-2r/a_0}$;

$$\frac{dD(r)}{dr} = Nr^2 \times \frac{-2}{a_0} e^{-2r/a_0} + 2Nr \times e^{-2r/a_0} = 2Ne^{-2r/a_0}\left(r - \frac{r^2}{a_0}\right).$$

(b) $D(r)$ displays a turning point when

$$\frac{dD(r)}{dr} = 2Ne^{-2r/a_0}\left(r - \frac{r^2}{a_0}\right) = 0,$$

i.e. when $r = \frac{r^2}{a_0} \Rightarrow r = a_0$. When $r = a_0$, $D(r) = Na_0^2 e^{-2}$.

(c) $\dfrac{d^2 D(r)}{dr^2} = -\dfrac{4}{a_0} Ne^{-2r/a_0}\left(r - \dfrac{r^2}{a_0}\right) + 2Ne^{-2r/a_0}\left(1 - \dfrac{2r}{a_0}\right)$

$$= 2Ne^{-2r/a_0}\left\{1 - \frac{2r}{a_0} - \frac{2r}{a_0} + \frac{2r^2}{a_0^2}\right\}.$$

$$\Rightarrow \frac{d^2 D(r)}{dr^2} = 2Ne^{-2r/a_0}\left\{1 - \frac{4r}{a_0} + \frac{2r^2}{a_0^2}\right\}.$$

When $r = a_0$, $\dfrac{d^2 D(r)}{dr^2} = 2Ne^{-2}\{1 - 4 + 2\}$

$$= -2Ne^{-2} \Rightarrow \text{maximum.}$$

(d)   Points of inflection occur when

$$\frac{d^2 D(r)}{dr^2} = 2Ne^{-2r/a_0}\left\{1 - \frac{4r}{a_0} + \frac{2r^2}{a_0^2}\right\} = 0,$$

i.e. when $1 - \dfrac{4r}{a_0} + \dfrac{2r^2}{a_0^2} = 0 \Rightarrow \dfrac{r}{a_0} = \dfrac{4 \pm \sqrt{16 - 4 \times 2 \times 1}}{4}$

$$= 1 \pm \frac{\sqrt{8}}{4} = 1 \pm \frac{\sqrt{2}}{2} = 1 \pm \frac{1}{\sqrt{2}},$$

i.e. when $r = \left(1 \pm \dfrac{1}{\sqrt{2}}\right)a_0.$

4.13. $p = \dfrac{nRT}{V}; \ \dfrac{\partial p}{\partial T} = \dfrac{nR}{V}; \ \dfrac{\partial p}{\partial V} = -\dfrac{nRT}{V^2}.$

Chapter 5

5.1. $y = x^{1/3}; \dfrac{dy}{dx} = \dfrac{1}{3}x^{-2/3} \Rightarrow dy = \dfrac{1}{3}x^{-2/3}dx.$

(a) $dy = \dfrac{1}{3} \times 27^{-2/3} \times 3 = 0.111; \Delta y = 30^{1/3} - 27^{1/3} = 0.107;$

    $\Rightarrow dy$ overestimates Δy by 3.6%.

(b) $dy = \dfrac{1}{3} \times 27^{-2/3} \times 0.1 = 0.0037037;$

    $\Delta y = 27.1^{1/3} - 27^{1/3} = 0.0036991;$

    $\Rightarrow dy$ overestimates Δy by 0.12%.

5.2. (a) $C_p = \alpha + \beta T + \gamma T^2$

$C_{p,500\,K} = 14.143 \text{ J K}^{-1}\text{ mol}^{-1} + 75.495 \times 10^{-3}\text{ J K}^{-2}\text{ mol}^{-1}$

       $\times 500\text{ K} - 179.64 \times 10^{-7}\text{ J K}^{-3}\text{ mol}^{-1} \times 500^2\text{ K}^2$

       $\Rightarrow C_{p,500\,K} = 47.3995\text{ J K}^{-1}\text{ mol}^{-1}.$

       Similarly, $C_{p,650\,K} = 55.625\text{ J K}^{-1}\text{ mol}^{-1}.$

(b) $dC_p = (\beta + 2\gamma T) \times dT = \left\{ \begin{array}{l} 75.495 \times 10^{-3}\text{ J K}^{-2}\text{mol}^{-1} \\ -(2 \times 179.64 \times 10^{-7}\text{ J K}^{-3}\text{mol}^{-1}) \end{array} \right\} \times 150\text{K}$

    $dC_p = 8.62965\text{ J K}^{-1}\text{mol}^{-1}$

       $\Rightarrow C_{p,650\text{K}} = (47.3995 + 8.62965)\text{ J K}^{-1}\text{mol}^{-1}$

       $= 56.029\text{ J K}^{-1}\text{mol}^{-1}$

(c)   The estimate is 0.73% larger than the actual value.

5.3. $z = xy/2; \; dz = \dfrac{\partial z}{\partial x}dx + \dfrac{\partial z}{\partial y}dy = \dfrac{y}{2}dx + \dfrac{x}{2}dy.$

5.4. (a) $U = f(V, T); \; dU = \left(\dfrac{\partial U}{\partial V}\right)_T dV + \left(\dfrac{\partial U}{\partial T}\right)_V dT.$

(b)   (i) $\pi_T = \left(\dfrac{\partial U}{\partial V}\right)_T; \; C_V = \left(\dfrac{\partial U}{\partial T}\right)_V.$

    (ii) $dU = \pi_T dV + C_V dT$

         $= (840\text{ J m}^{-3} \times -10^{-4}\text{ m}^3) + (27.32\text{ J K}^{-1} \times 2\text{ K})$

         $= 54.56\text{ J}.$

Contribution from compression is far smaller than that from the increase in temperature.

5.5. (a) $dV = \dfrac{\partial V}{\partial a}da + \dfrac{\partial V}{\partial b}db + \dfrac{\partial V}{\partial c}dc = bc \times da + ac \times db + ab \times dc.$

(b)   The relative error in V is $\dfrac{dV}{V}$ and so using the result from (a), we have:

$$\frac{dV}{V} = \frac{bc \times da + ac \times db + ab \times dc}{abc} \Rightarrow \frac{dV}{V} = \frac{da}{a} + \frac{db}{b} + \frac{dc}{c}$$

The percentage error in V is $100 \times \left(\dfrac{da}{a} + \dfrac{db}{b} + \dfrac{dc}{c} \right)$.

5.6. (a) $(CaCO_3) \times 4 = Ca_4C_4O_{12}$

$\Rightarrow$ molar mass $= 4 \times 40.08 + 4 \times 12.01 + 12 \times 16.00$

$= 400.36 \text{ g mol}^{-1} = 0.40036 \text{ kg mol}^{-1} = 6.648 \times 10^{-25} \text{ kg}.$

(b) $V = abc = 4.94 \times 10^{-10} \text{ m} \times 7.94 \times 10^{-10} \text{ m} \times 5.72 \times 10^{-10} \text{ m}$

$= 2.244 \times 10^{-28} \text{ m}^3.$

$$\rho = M/V \Rightarrow \rho = \frac{6.648 \times 10^{-25} \text{ kg}}{2.244 \times 10^{-28} \text{ m}^3} = 2.963 \times 10^3 \text{ kg m}^{-3}.$$

(c) $\dfrac{dV}{V} = \dfrac{da}{a} + \dfrac{db}{b} + \dfrac{dc}{c} = \dfrac{0.005 \times 10^{-10}}{4.94 \times 10^{-10}} + \dfrac{0.005 \times 10^{-10}}{7.94 \times 10^{-10}} + \dfrac{0.005 \times 10^{-10}}{5.72 \times 10^{-10}}$

$= 1.012 \times 10^{-3} + 6.28 \times 10^{-4} + 8.74 \times 10^{-4} = 2.51 \times 10^{-3}$

$\Rightarrow 0.25\%$ error.

(d)   Greatest unit cell volume $= 4.945 \times 10^{-10} \times 7.945 \times 10^{-10}$

$\times 5.725 \times 10^{-10} \text{ m}^3 = 2.249 \times 10^{-28} \text{ m}^3$

$\Rightarrow \rho = 2.956 \times 10^3 \text{ kg m}^{-3} \Rightarrow 0.237\%$ error.

Least unit cell volume $= 4.935 \times 10^{-10} \times 7.935 \times 10^{-10}$

$\times 5.715 \times 10^{-10} \text{ m}^3 = 2.238 \times 10^{-28} \text{ m}^3$

$\Rightarrow \rho = 2.971 \times 10^3 \text{ kg m}^{-3} \Rightarrow 0.27\%$ error.

Chapter 6

6.1. (a) $\dfrac{d}{dx}e^{2x} = 2e^{2x} \Rightarrow \displaystyle\int 2e^{2x} = 2\int e^{2x} = e^{2x} + B \Rightarrow \int e^{2x} = \dfrac{1}{2}e^{2x} + C.$

(b) $\dfrac{d}{dx}\dfrac{1}{1+e^x} = \dfrac{d}{dx}(1+e^x)^{-1}$; Let $u = 1 + e^x$; $\dfrac{du}{dx} = e^x$;

$\dfrac{d}{dx}(1+e^x)^{-1} = -(1+e^x)^{-2} \times e^x = -\dfrac{e^x}{(1+e^x)^2}$

Thus, it follows that $-\int \dfrac{e^x}{(1+e^x)^2}dx = \dfrac{1}{1+e^x} + B$ and so

$$\int \frac{e^x}{(1+e^x)^2}dx = -\frac{1}{1+e^x} + C.$$

6.2. $\displaystyle\int 9x^2dx + 2e^{2x} + \frac{1}{x}dx = \int 9x^2dx + \int 2e^{2x}dx + \int \frac{1}{x}dx$

$$= 3x^3 + e^{2x} + \ln x + C.$$

6.3. (a) $\displaystyle\int xe^{-x}dx = uv - \int v\frac{du}{dx}dx$, where $u = x$ and $\dfrac{dv}{dx} = e^{-x}$

$$\Rightarrow \int xe^{-x}dx = -xe^{-x} - \int -e^{-x} \times 1 \, dx = -xe^{-x} - e^{-x}$$

$$= -e^{-x}(x+1) + C.$$

(b) $\displaystyle\int xe^{-x}dx = uv - \int v\frac{du}{dx}dx$ where $u = e^{-x}$ and $\dfrac{dv}{dx} = x$

$$\Rightarrow \int xe^{-x}dx = e^{-x} \times \frac{1}{2}x^2 + \int \frac{1}{2}x^2 \times e^{-x}dx.$$

This solution requires us to integrate $\int \frac{1}{2}x^2 \times e^{-x}dx$, which is arguably more complicated than the original integral. Thus, method (a) would seem the more appropriate.

6.4. $\displaystyle\int xe^{ax^2}dx$. For $u = x^2$, $du = \dfrac{du}{dx}dx = 2x\,dx \Rightarrow dx = \dfrac{1}{2x}du$

$$\Rightarrow \int xe^{au}\frac{du}{2x} = \frac{1}{2}\int e^{au}du = \frac{1}{2a}e^{au} + C \Rightarrow \int xe^{ax^2}dx = \frac{1}{2a}e^{ax^2} + C.$$

6.5. $\displaystyle\int \frac{x}{(1-x^2)^{1/2}}dx$. For $u = 1-x^2, du = \dfrac{du}{dx}dx = -2x\,dx \Rightarrow dx = -\dfrac{1}{2x}du$

$$\Rightarrow \int \frac{-x}{u^{1/2}}\frac{1}{2x}du = -\int \frac{1}{2u^{1/2}}du = -\frac{1}{2}\int u^{-1/2}du = -\frac{1}{2}2u^{1/2} + C$$

$$= -u^{1/2} + C = -(1-x^2)^{1/2} + C.$$

6.6 (a) $\displaystyle\int x(x^2+4)^{1/2}dx$. For $u = x^2+4, du = \dfrac{du}{dx}dx = 2x\,dx$

$$\Rightarrow dx = \frac{1}{2x}du \Rightarrow \int xu^{1/2}\frac{1}{2x}du = \frac{1}{2}\int u^{1/2}du = \frac{1}{2}\frac{2}{3}u^{3/2} + C$$

$$= \frac{1}{3}u^{3/2} + C = \frac{1}{3}(x^2+4)^{3/2} + C.$$

(b) $\int \dfrac{1}{x\ln x}\,dx$. For $u = \ln x$, $du = \dfrac{du}{dx}dx = \dfrac{1}{x}dx \Rightarrow dx = x\,du$

$\Rightarrow \int \dfrac{x}{xu}\,du = \int \dfrac{1}{u}\,du = \ln u + C = \ln(\ln x) + C.$

6.7. (a) $\int \dfrac{x}{(1-x^2)^{1/2}}\,dx$. For $x = \cos u$, $\dfrac{dx}{du} = -\sin u \Rightarrow dx = -\sin u\,du$

$\Rightarrow \int \dfrac{-\cos u \times \sin u}{(1-\cos^2 u)^{1/2}}\,du = -\int \dfrac{\cos u \times \sin u}{(\sin^2 u)^{1/2}}\,du$

$= -\int \dfrac{\cos u \times \sin u}{(\sin u)}\,du = -\int \cos u\,du = -\sin u + C$

$= -(1-\cos^2 u)^{1/2} + C = -(1-x^2)^{1/2} + C.$

(b) $\int \dfrac{\cos x}{\sin x}\,dx$. For $u = \sin x$, $\dfrac{du}{dx} = \cos x \Rightarrow dx = \dfrac{du}{\cos x}$

$\Rightarrow \int \dfrac{\cos x}{u}\dfrac{du}{\cos x} = \int \dfrac{du}{u} = \ln u + C = \ln(\sin x) + C.$

6.8. (a) (i) $\int_1^2 \dfrac{1}{x^3}\,dx = \left[-\dfrac{1}{2}x^{-2} + C\right]_1^2 = -\dfrac{1}{2}2^{-2} + \dfrac{1}{2}1^{-2} = -\dfrac{1}{8} + \dfrac{1}{2} = \dfrac{3}{8}.$

(ii) $\int_0^2 x(x^2+4)^{1/2}\,dx = \left[\dfrac{1}{3}(x^2+4)^{3/2} + C\right]_0^2 = \dfrac{1}{3}8^{3/2} - \dfrac{1}{3}4^{3/2}$

$= \dfrac{1}{3}8\sqrt{8} - \dfrac{1}{3}4\sqrt{4} = \dfrac{16}{3}\sqrt{2} - \dfrac{8}{3} = 4.876.$

(b) $\int_0^2 \dfrac{x}{(x^2+4)}\,dx$. Let $u = x^2+4$; $\dfrac{du}{dx} = 2x \Rightarrow dx = \dfrac{du}{2x}$

$\int_0^2 \dfrac{x\,du}{u\,2x} = \dfrac{1}{2}\int_0^2 \dfrac{du}{u} = \dfrac{1}{2}\left[\ln(x^2+4)\right]_0^2 = \dfrac{1}{2}\{\ln(2^2+4) - \ln 4\}$

$= \dfrac{1}{2}(\ln 8 - \ln 4) = \dfrac{1}{2}\ln\dfrac{8}{4} = \dfrac{1}{2}\ln 2.$

6.9. $W = \int_{V_a}^{V_b} p\,dV;\ p = nRT/V \Rightarrow$

$W = \int_{V_a}^{V_b} \dfrac{nRT}{V}\,dV = nRT\int_{V_a}^{V_b}\dfrac{dV}{V} = nRT\,[\ln V]_{V_a}^{V_b} = nRT\{\ln V_b - \ln V_a\}$

$= nRT\ln\dfrac{V_b}{V_a}.$

6.10. (a) $\dfrac{d}{dT}\ln K=\dfrac{\Delta H^{\circ}}{RT^2}\Rightarrow\ln K=\displaystyle\int\dfrac{\Delta H^{\circ}}{RT^2}dT=\dfrac{\Delta H^{\circ}}{R}\int\dfrac{1}{T^2}dT=\dfrac{-\Delta H^{\circ}}{RT}+C.$

(b) $\quad\Delta\ln K=\displaystyle\int_{500\text{K}}^{600\text{K}}\dfrac{\Delta H^{\circ}}{RT^2}dT=\dfrac{-\Delta H^{\circ}}{R\times600}-\dfrac{-\Delta H^{\circ}}{R\times500}$

$$=\ \Delta H^{\circ}\left\{\dfrac{1}{R\times500}-\dfrac{1}{R\times600}\right\}$$

$$=42.3\times10^3\ \text{J mol}^{-1}\left\{\dfrac{1}{8.314\ \text{J K}^{-1}\ \text{mol}^{-1}\times500\ \text{K}}\right.$$

$$\left.-\dfrac{1}{8.314\ \text{J K}^{-1}\ \text{mol}^{-1}\times600\ \text{K}}\right\}=1.696.$$

Chapter 7

7.1. (a) $\ y=x^{-1};\quad\dfrac{dy}{dx}=-x^{-2}=-y^2;\quad$ 1st order differential equation.

$\dfrac{d^2y}{dx^2}=2x^{-3}=2y^3;$ 2nd order.

(b) $\ y=\cos ax;\dfrac{dy}{dx}=-a\sin ax;\dfrac{d^2y}{dx^2}=-a^2\cos ax=-a^2y.$

(c) $\ y=Ae^{4x};\dfrac{d}{dx}Ae^{4x}-4Ae^{4x}=4Ae^{4x}-4Ae^{4x}=0$

$\dfrac{d^2}{dx^2}Ae^{4x}-5\dfrac{d}{dx}Ae^{4x}+4Ae^{4x}=16Ae^{4x}-5\times4Ae^{4x}+4Ae^{4x}=0.$

7.2. $\dfrac{dy}{dx}=-6y^2\Rightarrow\dfrac{dy}{y^2}=-6\,dx\Rightarrow\displaystyle\int\dfrac{dy}{y^2}=\int-6\,dx\Rightarrow$

$$\dfrac{-1}{y}=-6x+C\Rightarrow1=y(6x+D)\Rightarrow y=\dfrac{1}{6x+D}$$

$$y=1,x=0,\ \text{so }1=\dfrac{1}{D}\Rightarrow D=1.\ \text{Therefore }y=\dfrac{1}{6x+1}.$$

7.3. $\dfrac{dy}{dx}=-\lambda y\Rightarrow\dfrac{dy}{y}=-\lambda\,dx\Rightarrow\displaystyle\int\dfrac{dy}{y}=\int-\lambda\,dx\Rightarrow\ln y=-\lambda x+C.$

$$y=N,\ x=0,\ \text{so }\ln N=C\Rightarrow\ln y=-\lambda x+\ln N\Rightarrow\ln\dfrac{y}{N}=-\lambda x$$

$$\dfrac{y}{N}=e^{-\lambda x}\Rightarrow y=Ne^{-\lambda x}.$$

7.4. (a) $\dfrac{d}{dx}\left(\dfrac{1}{3}e^x + Ce^{-2x}\right) + 2\left(\dfrac{1}{3}e^x + Ce^{-2x}\right) = \dfrac{1}{3}e^x - 2Ce^{-2x}$

$\quad + \dfrac{2}{3}e^x + 2Ce^{-2x} = e^x.$

(b) $\dfrac{dy}{dx} + \dfrac{y}{x} = x^2; P(x) = \dfrac{1}{x}; Q(x) = x^2;$

$\quad R(x) = e^{\int \frac{1}{x}dx} = e^{\ln x + C} = e^C e^{\ln x} = e^C x = Ax.$

Here $g(x) = \ln x$. Therefore $y = \dfrac{1}{x}\displaystyle\int x \times x^2 dx$

$\quad = \dfrac{1}{x} \times \left\{\dfrac{x^4}{4} + C\right\} = \dfrac{x^3}{4} + \dfrac{C}{x}$

$\quad y = 0, x = 1;$ so $0 = \dfrac{1}{4} + C$

$\quad \Rightarrow C = -\dfrac{1}{4} \quad y = \dfrac{x^3}{4} - \dfrac{1}{4x}.$

7.5. (a) $P(x) \equiv \lambda_2$ and $Q(x) \equiv \lambda_1(N_1)_0 e^{-\lambda_1 t}.$

$\quad R(t) = e^{\int \lambda_2 dt} = e^{\lambda_2 t + C} = Ae^{\lambda_2 t},$ where $\lambda_2 t = g(t).$

(b) $N_2 = e^{-\lambda_2 t}\displaystyle\int e^{\lambda_2 t}\lambda_1(N_1)_0 e^{-\lambda_1 t} dt = e^{-\lambda_2 t}\displaystyle\int \lambda_1(N_1)_0 e^{(\lambda_2 - \lambda_1)t} dt$

$\quad = e^{-\lambda_2 t}\left(\dfrac{\lambda_1(N_1)_0}{\lambda_2 - \lambda_1} e^{(\lambda_2 - \lambda_1)t} + C\right) = \dfrac{\lambda_1(N_1)_0}{\lambda_2 - \lambda_1} e^{(\lambda_2 - \lambda_2 - \lambda_1)t} + Ce^{-\lambda_2 t}$

$\quad = \dfrac{\lambda_1(N_1)_0}{\lambda_2 - \lambda_1} e^{-\lambda_1 t} + Ce^{-\lambda_2 t}.$

(c) If $N_2 = 0$ at $t = 0$, then $0 = \dfrac{\lambda_1(N_1)_0}{\lambda_2 - \lambda_1} + C$

$\quad \Rightarrow C = -\dfrac{\lambda_1(N_1)_0}{\lambda_2 - \lambda_1}.$ Therefore $N_2 = \dfrac{\lambda_1(N_1)_0}{\lambda_2 - \lambda_1} e^{-\lambda_1 t}$

$\quad - \dfrac{\lambda_1(N_1)_0}{\lambda_2 - \lambda_1} e^{-\lambda_2 t} = \dfrac{\lambda_1(N_1)_0}{\lambda_2 - \lambda_1}(e^{-\lambda_1 t} - e^{-\lambda_2 t}).$

(d) $\dfrac{dN_1}{dt} = -\lambda_1 N_1$ has same form as $\dfrac{dy}{dx} = -\lambda y,$ for which

the solution is $y = Ne^{-\lambda x}$ for $y = N, x = 0.$

Thus it follows that the solution to $\frac{dN_1}{dt} = -\lambda_1 N_1,$ where $N_1 = (N_1)_0,$ $t = 0,$ is given by $N_1 = (N_1)_0\, e^{-\lambda_1 t} = e^{-\lambda_1 t},$ since $(N_1)_0 = 1$ mol.

(e) Since $N_1 + N_2 + N_3 = 1$ mol and $N_1 = e^{-\lambda_1 t}$, $N_2 = \frac{\lambda_1}{\lambda_2 - \lambda_1}(e^{-\lambda_1 t} - e^{-\lambda_2 t})$, it follows that $N_3 = 1 - e^{-\lambda_1 t} - \frac{\lambda_1}{\lambda_2 - \lambda_1}(e^{-\lambda_1 t} - e^{-\lambda_2 t})$

$$\Rightarrow N_3 = 1 - e^{-\lambda_1 t} - \frac{\lambda_1}{\lambda_2 - \lambda_1}e^{-\lambda_1 t} + \frac{\lambda_1}{\lambda_2 - \lambda_1}e^{-\lambda_2 t}$$

$$= 1 - \frac{\lambda_2 - \lambda_1}{\lambda_2 - \lambda_1}e^{-\lambda_1 t} - \frac{\lambda_1}{\lambda_2 - \lambda_1}e^{-\lambda_1 t} + \frac{\lambda_1}{\lambda_2 - \lambda_1}e^{-\lambda_2 t}$$

$$= 1 - \left(\frac{\lambda_2 - \lambda_1}{\lambda_2 - \lambda_1}e^{-\lambda_1 t} + \frac{\lambda_1}{\lambda_2 - \lambda_1}e^{-\lambda_1 t} - \frac{\lambda_1}{\lambda_2 - \lambda_1}e^{-\lambda_2 t}\right)$$

$$= 1 - \left(\frac{(\lambda_2 - \lambda_1 + \lambda_1)e^{-\lambda_1 t} - \lambda_1 e^{-\lambda_2 t}}{\lambda_2 - \lambda_1}\right)$$

$$= 1 - \left(\frac{\lambda_2 e^{-\lambda_1 t} - \lambda_1 e^{-\lambda_2 t}}{\lambda_2 - \lambda_1}\right)$$

$$= 1 + \left(\frac{\lambda_1 e^{-\lambda_2 t} - \lambda_2 e^{-\lambda_1 t}}{\lambda_2 - \lambda_1}\right).$$

(f) $t_{1/2} = \frac{\ln 2}{\lambda_i}$

Thus, $\lambda_1 = \frac{\ln 2}{t_{1/2}} = \frac{\ln 2}{23.5}$ min^{-1} = 2.95×10^{-2} min^{-1}.

$$\lambda_2 = \frac{\ln 2}{t_{1/2}} = \frac{\ln 2}{2.3}\text{ day}^{-1} = \frac{\ln 2}{3312}\text{ min}^{-1} = 2.093 \times 10^{-4}\text{ min}^{-1}.$$

(g) $N_2 = \frac{\lambda_1}{\lambda_2 - \lambda_1}(e^{-\lambda_1 t} - e^{-\lambda_2 t})$: maximum when $\frac{dN_2}{dt} = 0$

$$\frac{dN_2}{dt} = \frac{\lambda_1}{\lambda_2 - \lambda_1}(-\lambda_1 e^{-\lambda_1 t} - -\lambda_2 e^{-\lambda_2 t}) = \frac{\lambda_1}{\lambda_2 - \lambda_1}(\lambda_2 e^{-\lambda_2 t} - \lambda_1 e^{-\lambda_1 t}) = 0$$

when $\lambda_2 e^{-\lambda_2 t} = \lambda_1 e^{-\lambda_1 t}$.

Taking logs gives $\ln \lambda_2 - \lambda_2 t = \ln \lambda_1 - \lambda_1 t$

$$\Rightarrow \ln\frac{\lambda_2}{\lambda_1} = \lambda_2 t - \lambda_1 t \Rightarrow \ln\frac{\lambda_2}{\lambda_1} = \lambda_2 t - \lambda_1 t$$

$$\Rightarrow t = \frac{1}{\lambda_2 - \lambda_1}\ln\frac{\lambda_2}{\lambda_1} = \frac{1}{0.0002092 - 0.0295}\ln\frac{0.0002092}{0.0295} = 168.96\text{ min.}$$

Thus, $N_{2,\text{max}} = \frac{0.0295}{0.0002092 - 0.0295}(e^{-0.0295 \times 168.96} - e^{-0.0002092 \times 168.96})$

$= 0.965$ mol.

7.6. (a)

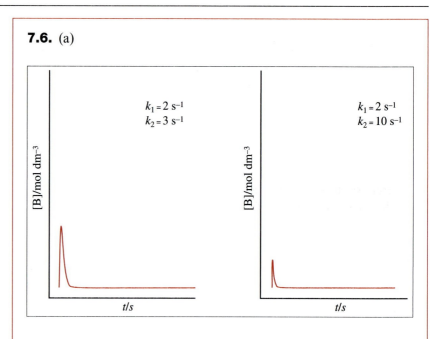

(b)   As k_2 increases relative to k_1, so the maximum concentration of the intermediate decreases.

7.7. (a) $\frac{d^2 y}{dx^2} = n^2 y \Rightarrow y = A e^{nx} + B e^{-nx}$.

(b) $\sinh nx = \frac{1}{2}(e^{nx} - e^{-nx}) \Rightarrow e^{nx} - e^{-nx} = 2 \sinh nx.$

$\cosh nx = \frac{1}{2}(e^{nx} + e^{-nx}) \Rightarrow e^{nx} + e^{-nx} = 2 \cosh nx$

$\Rightarrow 2 e^{nx} = 2 \sinh nx + 2 \cosh nx \Rightarrow e^{nx} = \sinh nx + \cosh nx$

$\Rightarrow 2 e^{-nx} = 2 \cosh nx - 2 \sinh nx \Rightarrow e^{-nx} = \cosh nx - \sinh nx$

$\Rightarrow y = A \sinh nx + A \cosh nx + B \cosh nx - B \sinh nx$

$y = (A - B) \sinh nx + (A + B) \cosh nx.$

(c) $0 = (A - B) \sinh 0 + (A + B) \cosh 0 = A + B \Rightarrow B = -A$

$y = (A - -A) \sinh nx + (A + -A) \cosh nx = 2A \sinh nx.$

7.8. $\hat{D}^2 (A \cos nx + B \sin nx) = -A n^2 \cos nx - B n^2 \sin nx$

$$= -n^2 (A \cos nx + B \sin nx)$$

7.9. $\frac{d^2 y}{dx^2} - 5 \frac{dy}{dx} + 6y = 0$ has the same form as $\frac{d^2 y}{dx^2} + c_1 \frac{dy}{dx} = -c_2 y,$

where $c_1 = -5$ and $c_2 - 6.$

Factorizing yields $(\hat{D}^2 - 5\hat{D})y = -6y$, where the operator $\hat{A} = \hat{D}^2 - 5\hat{D}$ and the eigenvalue $\lambda = -6$. We know from Worked Problem 7.6 that e^{nx} is an eigenfunction of the operators $\hat{D}^2$ and $\hat{D}$ and so we need to find the appropriate values of n that are consistent with the operator, $\hat{A} = \hat{D}^2 - 5\hat{D}$ and an eigenvalue $\lambda = -6$. Thus:

$$\hat{A}e^{nx} = (n^2 - 5n)e^{nx} = -6e^{nx} \Rightarrow (n^2 - 5n) = -6 \Rightarrow n^2 - 5n + 6 = 0$$

Factorizing yields $(n - 2)(n - 3) = 0$, and so $n = 2, 3$. Thus the general solution is $y = Be^{3x} + Ce^{2x}$. Imposing the boundary condition $y = 0$, $x = 0 \Rightarrow 0 = B + C$ and $dy/dx = 1$ when $x = 0$ $\Rightarrow \frac{dy}{dx} = 3Be^{3x} + 2Ce^{2x} = 1$, when $x = 0$. Thus:
$3B + 2C = 1$. However, $B = -C$, and so $-3C + 2C = 1 \Rightarrow C = -1$ and so $B = 1$. Therefore the solution is $y = e^{3x} - e^{2x}$.

Subject Index